# Urban Expansion and Food Security in New Zealand

This book examines suburban development in New Zealand and its conflict with and impact on local horticulture and food security.

Drawing on an ethnographic study of Auckland's rapidly expanding urban periphery, combined with comparative case studies from California in the USA and Victoria in Australia, this book examines how the profit-making strategies of property developers and landowners drastically reshape work and life at the edge of cities. With a significant portion of the world's croplands lying adjacent to cities, the accelerating pace of urban sprawl across the planet places unprecedented pressure on the productivity and even existence of these vital food bowl regions. This book examines how the demand for more land for development at the urban periphery collides with concerns over local food security and the protection of ecosystem services. It analyses land use policy, historical records, and physical patterns of development, alongside participant observation of local events. It combines this with interviews with government officials, property developers, landowners, local residents, and horticulturists. By combining these narratives of the hectic and lucrative business of suburban property development with the collapse of local horticulture, this book shows how the realignment of New Zealand's interests of financial profitability over other concerns led to the transformation of urban peripheries from a productive food bowl to an investment vehicle.

This book will be of great interest to students and scholars of urban food and agriculture, urban planning and development, and rural–urban studies.

**Benjamin Felix Richardson** is a doctoral candidate and a professional teaching fellow in the Department of Anthropology at the University of Auckland, New Zealand.

## Routledge Studies in Food, Society and the Environment

**The Vegan Evolution**
Transforming Diets and Agriculture
*Gregory F. Tague*

**Food Loss and Waste Policy**
From Theory to Practice
*Edited by Simone Busetti and Noemi Pace*

**Rewilding Food and the Self**
Critical Conversations from Europe
*Edited by Tristan Fournier and Sébastien Dalgalarrondo*

**Critical Mapping for Sustainable Food Design**
Food Security, Equity and Justice
*Audrey G. Bennett and Jennifer A. Vokoun*

**Community Food Initiatives**
A Critical Reparative Approach
*Edited by Oona Morrow, Esther Veen, and Stefan Wahlen*

**Food Futures in Education and Society**
*Edited by Gurpinder Singh Lalli, Angela Turner, and Marion Rutland*

**The Soybean Through World History**
Lessons for Sustainable Agrofood Systems
*Matilda Baraibar Norberg and Lisa Deutsch*

**Urban Expansion and Food Security in New Zealand**
The Collapse of Local Horticulture
*Benjamin Felix Richardson*

For more information about this series, please visit: www.routledge.com/books/series/RSFSE

# Urban Expansion and Food Security in New Zealand

## The Collapse of Local Horticulture

**Benjamin Felix Richardson**

Routledge
Taylor & Francis Group
LONDON AND NEW YORK

earthscan
from Routledge

First published 2024
by Routledge
4 Park Square, Milton Park, Abingdon, Oxon OX14 4RN

and by Routledge
605 Third Avenue, New York, NY 10158

*Routledge is an imprint of the Taylor & Francis Group, an informa business*



*British Library Cataloguing-in-Publication Data*
A catalogue record for this book is available from the British Library

*Library of Congress Cataloging-in-Publication Data*
Names: Richardson, Benjamin Felix, author.
Title: Urban expansion and food security in New Zealand : the collapse of local horticulture / Benjamin Felix Richardson.
Description: New York, NY : Routledge, 2024 |
Includes bibliographical references and index.
Identifiers: LCCN 2023010267 (print) | LCCN 2023010268 (ebook) |
ISBN 9781032504223 (hardback) | ISBN 9781032509334 (paperback) |
ISBN 9781003400325 (ebook)
Subjects: LCSH: Food security—New Zealand. | Rural-urban relations—New Zealand. | Land use, Urban—New Zealand. |
Urban ecology (Sociology)—New Zealand.
Classification: LCC HT384.N45 R53 2023 (print) |
LCC HT384.N45 (ebook) | DDC 307.740993—dc23/eng/20230426
LC record available at https://lccn.loc.gov/2023010267
LC ebook record available at https://lccn.loc.gov/2023010268

ISBN: 978-1-032-50422-3 (hbk)
ISBN: 978-1-032-50933-4 (pbk)
ISBN: 978-1-003-40032-5 (ebk)

DOI: 10.4324/9781003400325

Typeset in Times New Roman
by codeMantra

# Contents

# Illustrations

## Figures

## Tables

# Acknowledgements

This book would not exist without the guidance of my supervisors at the University of Auckland. Mark Busse and Emma Sharp gave their time and their energy towards the creation of the thesis on which this book is based, over a period when both resources were in very short supply.

My research in Kumeu-Huapai could not have gone far without introductions, advice, and explanations from people who have made it their lives' work to understand the history of the area and its possible futures. These include past and present members of the Ngā Maunga Whakahii o Kaipara Development Trust including Margaret Kawharu and Shona Oliver; the local historians Derry Faulder, Chris Smellie, and Deborah Dunsford; and the members of both the Kumeu Community Action group and the Kumeu Agricultural and Horticultural Society.

The cooperation of many professionals working in government and in property development across Auckland's northwest was no less important. Their anonymous contributions have been vital to examining the conflict over land at the edge of the city.

I am very thankful to have friends and family who were supportive of the years spent conducting this research and who offered the feedback, counterarguments, and points of view necessary to the final product of both thesis and book. Special gratitude to my flatmates Alastair Harris, Stormy Taurere, Kelly Wootton, and Thomas Cain for having to live with me while I worked on this and to Judith Glasson for everything.

# Introduction

## The suburban frontier

As the world's cities grow, fortunes are made from the land that lies in their path. Across the planet, people are moving from rural areas to urban centres at an unprecedented rate, but rural land is being converted to urban uses at an even faster pace. As the urban population doubles between 2000 and 2040, the sheer area of urban land is on track to triple (d'Amour et al., 2017, p. 8939). Cities tend to be founded close to the planet's most productive agricultural areas for the simple reason that these significant sources of food are needed to support urban life in the first place. Because of this, some two-thirds of the world's croplands are adjacent to cities (*ibid.*). When demand for peri-urban[1] land rises, the new suburban and exurban communities of the city are built on land that once grew the food that sustained its people. This demand is increasingly driven by financial motives: real estate in cities and their peripheries is tied into global systems of finance as a highly valued investment of last resort in an unstable market (Waldron, 2019), and this instability itself has only been amplified by a marked rise in inequality and the concentration of capital in recent decades (Goda et al., 2017). Due to the convergence of all of these factors, real estate values are climbing rapidly worldwide (Joint Centre for Housing Studies of Harvard University, 2021) and parcels of formerly rural land at the edge of urban expansion are extremely lucrative when subdivided and converted into suburban housing or lifestyle living blocks, then sold to urban residents who can afford to take on the mortgage. In towns where working life was once defined by the provision of agricultural produce to a nearby city, the business of finance-driven property development is transforming the landscape and providing landowners and development professionals with unbeatable profitability.

Auckland City is the largest urban centre in New Zealand by far. Some 1.7 million people, a third of the country's entire population, live in the wider Auckland region (Auckland Council, 2018). By 2050, the area is forecast to be home to 2.4 million people (p. 9), 40 per cent of New Zealand's population at that time (*ibid.*). Auckland is also a major food bowl for the country, stocking supermarkets in the neighbouring regions of Northland and Waikato while providing vegetables to New Zealand's South Island in winter months when local supply cannot keep up (Deloitte, 2019, p. 26; Radio New Zealand, 2019). Today, more than a third of Auckland's most productive land has been subdivided into smaller blocks and then developed (Curran-Cournane et al., 2021, p. 3). This expansion of the suburban

DOI: 10.4324/9781003400325-1

frontier into rural Auckland has increased by nearly 30 per cent since the new millennium (p. 6). As growing land at the urban periphery becomes far more valuable as an investment vehicle than for producing food, policymakers and academics are concerned about the long-term costs of Auckland's sprawl and seek ways to reconcile the unending demand for land to be released for urbanization with the need to preserve food production and vital ecosystem services.

In my previous work, I investigated this phenomenon specifically in regard to Auckland's remaining food-producing areas in Pukekohe, south of Auckland City. Published material from this research (Richardson, 2021) focussed on how the financial success of suburban development in Pukekohe threatened to erode Auckland's food security, raising vegetable prices by up to 50 per cent across the country (Deloitte, 2019, p. 8) at a time when the viability of global food supply chains is in doubt. This threat, I argued, comes from the dominant valuations of land in Aotearoa that privilege financial returns over all other aspects of human life. In this book, however, I look at the business of suburban property development itself, its victory over local horticulture as the biggest moneymaker in Auckland's formerly rural towns, and how this economic transformation affects the lives of peri-urban residents. To investigate this, I sought out a peri-urban town in the process of suburbanization, a place where, unlike Pukekohe, horticultural production had been successfully dismantled in favour of land speculation and development within recent memory. I came up with Kumeu-Huapai, population 4,000 and rising fast, 20 km west of Auckland City, midway through a radical transformation from fruit orchards and stockyards to suburban homes and traffic jams. It is also only a few kilometres from where my parent's home stands and where some old friends yet live. Kumeu-Huapai's conversion into part of Auckland City occurred during the course of my own life and (in the times when I was living in Auckland) before my own eyes.

## Insider knowledge, outside context: ethnography in a familiar place

Looking back at his own research on the new urban-oriented elite in the English Home Counties where he grew up, Jesse Heley (2011) pointed to some unique advantages of "backyard ethnography". Noting that for ethnographers, being an "insider" or an "outsider" exists not on a single spectrum of commonality or "insider knowledge" shared with their interlocutors, but several axes of identity and understanding, Heley specifically argued that geographical familiarity is both an understudied aspect of researcher positionality (p. 222) and a strength for the early stages of ethnographic work (p. 223). By having some prior understanding of the actual physical landscape in which their interlocutors live and work, the researcher can have significant advantages in identifying some of the issues and conflicts that dominate life on the site, making introductions to interlocutors, and establishing their relationship with organizations in the area, "It follows that being local, in the sense of prolonged geographical proximity to a particular culture, helped me to identify suitable people and places for research" (p. 225). For precisely this reason, Kumeu-Huapai was more attractive to me of Auckland's new suburbs to investigate for this project. The town is also a good candidate for investigating the business of peri-urban development in Auckland due to its status as one of the fastest-growing frontiers of the city (Stats NZ, 2022) and the place with the highest rises in property values in the entire country between 2014 and 2019 (Oneroof, 2020). This aside, however, my passing familiarity with how the place has actually changed and grown in recent years – which fields have become expensive suburban houses and which businesses have occupied Kumeu-Huapai's main street – provides a very useful point of entry into my fieldwork.

At the same time, this literal geographical proximity to the place where one conducts ethnographic work creates new challenges, problems that are not nearly so prominent when one finds themselves in an unfamiliar place. Heley acknowledged a need to articulate and interrogate one's existing relationship with the site, to attempt to treat the locals of these places as either interlocutors for the purpose of research or as personal acquaintances, and not to mix these categories as far as is possible, but continued to encounter people with connections to his own life and family throughout his fieldwork, and noted that these connections further helped him gain access to the social lives of his subjects (Heley, 2011, p. 229). There were very few people I knew personally who continued to live in Kumeu-Huapai, and I made a point of not interviewing those I did know about development in the areas, seeking that separation. Most days when I conducted interviews and observed events, I, as Heley put it, "seldom felt more than a tourist and was never set to become a fixture that was missed" (p. 230). Yet, the very first person I interviewed – Derry Faulder, a local historian and long-standing member of the local Agricultural and Horticultural Society – turned out to have trained as an electrician's apprentice under my maternal grandfather and visited him in hospital before he died about half a century ago when my mother was very young. Moments like this were rare during my year of fieldwork, but these connections helped me place myself in context to some interlocutors and complicated my own understanding of a town I usually considered to have little bearing on my own life.

My ethnographic research for this project was conducted from March 2021 to April 2022. I drove the 20 km from Auckland City out to Kumeu-Huapai most days of the week to speak to interlocutors and visit ongoing sites of development. To see how the collapse of local horticulture and the rise of suburban property development in Kumeu-Huapai affected work and life in this urban periphery, I initially focussed on interviewing current and former participants in these two industries, as well as the planners, policymakers, and residents who were politically involved in conflicts over Kumeu-Huapai's future. Starting from the sites of new suburban developments in the town and from news articles or legal battles over development here at Auckland's northwestern frontier, I sought the professionals and advocates involved on all sides for unstructured interviews. Speaking to real estate agents, property developers, landowners, horticulturalists (current and former), local officials, and city planners, I was able to discuss the arguments of one party with another and draw out the interlocutor's own theories on the conflicts at the heart of peri-urban sprawl. To accompany these interviews and find new areas of investigation in my ethnographic research, I also attended public events and community meetings in Kumeu-Huapai.

These included the annual local Agricultural and Pastoral show, public forums or protests over infrastructure challenges and development policy, and a nationally significant (and for local landowners somewhat divisive) Hot Rod and Classic Car festival. Lastly, my observation work entailed attending open homes and auctions for new suburban houses in Kumeu-Huapai, as well as visits to the building sites of these soon-to-be family homes, in order to witness the transformation of a plot of rural land into a suburban street in a substantive manner. Through this combination of approaches, I sought to show how the rise of suburban development and the collapse of local horticulture is transforming life at the urban periphery.

This ethnographic work informed, and was in turn informed by, textual and archival research into the historic process of Auckland's urban sprawl. Aerial photography of the region from 1940 onwards was collated and stitched together to provide a time lapse of Kumeu-Huapai's transformation alongside census records. Court records and zoning change applications were examined to find relevant individuals and points of contention that required investigation, and policy documents from both local and central government authorities regarding urban development were compared to locate notable conflicts. To provide points of international comparison, this desk research also entailed two textual case studies of peri-urban development: the Australian city of Melbourne's unstable exurban frontier in the Macedon Ranges, and Fresno, California's profitable and deeply unequal suburban boom. These two examples share Auckland's legacy of colonial expropriation, rapid urbanization, and land speculation, but possess key differences in scale, ecological context, and political economic arrangements that offer useful contrasts to my ethnographic case study. By focussing the extent of my fieldwork on a single town at the edge of a growing city while drawing on existing literature for the same phenomenon in other parts of Auckland and across the world, I can present a detailed, substantive view of the forces transforming life in a formerly rural settlement and then use the observations made there to discuss the planetary phenomenon of accelerating suburbanization.

A combination of "backyard ethnography" in a familiar town and an analysis of financialized urban development in the world's wealthiest nations, the research that went into this book sought to apply the observations gained from the transformation of life in the Kumeu River Valley to a rapidly urbanizing and increasingly unequal world. As with any choice made during research, there are limitations to the approaches I have taken, as well as advantages. These are worked through as they appear in the course of the argument being developed, rather than presented in the introduction as a disclaimer, and then left aside for the remainder of the book. In seeking to present a picture of the very profitable business of peri-urban development at a time when real estate is deeply tied into the global financial system, I focussed on the testimony and the theories presented by professionals in property development, horticulture, and local government, as well as the narratives of long-time residents and local historians. It is my hope only that the perspectives presented here provide a useful insight into the forces that are driving the unprecedented transformation of rural land at the edge of fast-growing cities across the planet today.

**Who is doing the financializing?**

Because the research that led to this book is so preoccupied with the experiences of landowners and professionals in the property sector in a growing city, the process of financializing peri-urban land – and how this process interacts with our social lives and our ecosystem – is the overarching concern of every single chapter to come. Financialization is a very broad label, referring to a number of different ways in which economic, social, and political life is reshaped in service of seeking maximal financial profitability (Mader et al., 2020, p. 7). First used to describe the rising power of the finance sector over the USA's domestic economy in the late 1980s and early 1990s, the term has become exponentially more popular in the years following the 2007–8 Global Financial Crisis as a way to understand how financial systems can continue to exert power over the rest of human existence even when they "fail" so dramatically (Gudeman, 2016, p. 159; Mader et al., 2020, p. 3). Academic depictions of the financialization of life are often accompanied by examinations of Neoliberalism – a similarly wide definition aiming to describe forms of governance that have abandoned the post-war mid-century compromises of the welfare state and domestic production in favour of enabling financial extraction. The geographer David Harvey (2007), the sociologist Pierre Bourdieu (1998), and the anthropologist David Graeber (2009) each argued from their own respective positions that these two interrelated concepts – Financialization and Neoliberalism – represent a new system of global imperialism and exploitation searching for frontiers to capitalize on. The people who make their living from land at Auckland's urban periphery, whether from subdividing and selling it or by growing food on it, do so in this context of financial profitability overtaking all other governing concerns across the planet.

To move beyond this extremely broad view of a collection of systems and towards an ethnographic depiction of how individual agents of urban development actually understand and approach financializing land, two aspects of financialization will be examined in close detail in this book. The first is viewing it through the lens of a historical process occurring in a single place, a transformation of land uses and social lives towards the aim of profitability for a few interested parties. This investigation will also demonstrate that the process that we call Financialization today can historically be better conceptualized as a series of advances and retreats in financial power (Fasianos et al., 2018) going back at least as far as the initial European colonization of Aotearoa in the 19th century. This approach, however, will only be a major focus for the first two chapters and is sporadically returned to later in the book when discussing other case studies beyond that of the Kumeu River Valley. The second and more prominent lens that I will employ throughout this book is that of Financialization as a cultural practice (Aitken, 2020), something that individuals reproduce and react to in the ways that are available to them. The scope of options available to these subjects depends in turn on their relationships to the business of suburban development itself, the individual's position within what Bourdieu called the "territorial field" of a suburb in the making (2005, p. 130). Landowners with connections to large land development companies and officials in local or central government possess a very real capacity to transform their land into something that produces an unprecedented amount of profitability for themselves, but this knowledge is not evenly distributed. Landowners, public officials, and professionals in the development sector engage in communication and competition in order to change the land at the edge of the city in ways that most benefit them, which, for many individuals in this territorial field, involves the subsumption of all other concerns in the pursuit of financial success.

By spending a significant amount of time on the historical process of financialization in a given place and how the people of this place relate to practices of financially driven suburban development, I aim to demonstrate how human life and the ecosystem are transformed in service of financial profitability for a small number of individuals. The broad and flexible labels of Financialization and Neoliberalism possess a particularly strong explanatory power for those attempting to understand how financial extraction is becoming an increasingly powerful force across the planet today, but here I intend to show who exactly is taking part in the process of financialization rather than sketching out a vast and impersonal system. In formerly rural towns at the edge of rapidly growing cities, unmatched opportunities for converting land that once grew food into suburban housing exist due to a political economic landscape that encourages speculation on urban sprawl. Yet, within this context, any given plot of land can only become a brand-new suburb of more than a hundred single-family houses with the involvement of individuals who possess direct connections to the local area and a clear understanding of how to effectively financialize peri-urban land.

## This book

My overarching argument is thus: the business of peri-urban property development is massively and uniquely profitable for some of its participants precisely because of the costs that suburbanization inflicts on people and the land. The acceleration of suburban sprawl is not just a very prominent and visible example of "creative destruction" (Harvey, 2007) in the name of creating profitability but also one of the most immediate challenges faced by policymakers and planners across the world seeking to mitigate inequality and adapt to a changing climate. After this introductory chapter, the book is comprised of eight main chapters and then a brief conclusion. These eight chapters focus on different aspects of the phenomenon of suburban sprawl and have been divided into two parts. Chapters 1–4 focus on urban development in Kumeu-Huapai: how it works, its effect on life in the area, its history since the advent of European colonial expansion into Aotearoa, its effect on how land is valued, and its relationship to the broader political economy of Auckland's urban development. Following this, Chapters 5–8 bring in comparative cases from elsewhere in Auckland, in New Zealand, and the two international cases from Outer Melbourne and the California Central Valley. These chapters take the arguments made about the business of peri-urban development in Kumeu-Huapai in the first half of the book and place them into new regional, national, and global contexts. From these comparisons, the latter chapters examine the relationship that accelerating suburban sprawl has with class politics, local food security, and adaption to climate change the world over. By beginning with observations of a town of a few thousand at the edge of a city before moving to theories on global phenomena, I can take the time to develop a depiction of how urban sprawl is experienced at the levels of individuals and within a community before embarking on the much broader and more ambitious task of relating these insights to cities as a general concept.

In order to set the scene and provide context to readers who may not be familiar with colonial suburbia, Chapter 1 begins with an experiential section on the work of a project manager at Auckland's northwestern suburban frontier, detailing the actual sequence of processes and forms of labour and materials involved to make one single-family home. This frantic and rapid work of building new developments was preceded by the quiet disappearance of the productive rural landscape from the area, and this contrast between the fates of the two industries leads me to one of the first foundational points of this book: the transformation of Kumeu-Huapai's land – as well as the working lives of its residents – is an explicitly financial project, an intentional reorientation away from state-supported production and towards maximum profitability. Chapter 2 provides a historical context for this process of financialization, detailing the history of land use in the Kumeu River Valley since the arrival of European colonial interests in the region. From the expropriation of the land from Tangata Whenua[2] by large financial speculators, to its deforestation, then subdivision and sale to hopeful new horticulturalists (organized by Auckland-based professionals), to Kumeu-Huapai's mid-century florescence as an agricultural centre supplying produce to Auckland, and to the collapse of this productivity and meteoric rise of property development seen in recent memory, I show that the dominance of financial motives towards the Kumeu River Valley is not all a new phenomenon but a core part of the lands colonial history, only held back in the middle of the 20th century by a state interest in local production.

With this dynamic – a shift in favour of financialized urban development as a result of the realignment of state priorities – in mind, Chapters 3 and 4 explore how the business of peri-urban development requires transformations of legal and political processes, as well as systems of values, in order to succeed. Chapter 3 looks specifically at the axiologies – value systems – used for land in the Kumeu River Valley and elsewhere at Auckland's suburban frontier. I argue that individuals with an interest in peri-urban development have worked to redefine what "productive land" is to allow rezoning and fragmentation of rural land. These redefinitions have created a vicious cycle of suburban expansion and horticultural collapse, as horticulture operations are classed as unproductive and subdivided leading to neighbouring producers struggling with their work and likewise selling their land. I provide a possible remedy to this issue in the form of Hicks and Curran-Cournane's revised "Farm-Scale" Land Use Capability model (2017) that can help us better understand the value of growing land. Chapter 4 continues to look at legal and political responses to sprawl to show how this expansion of finance-led development ties into Auckland City's political economy. Arguing against a prominent body of literature that has identified land supply as the culprit for skyrocketing housing unaffordability across the planet, I point to data from new suburban developments that show how attempts to free up land for development only increase house prices by making the sector a more attractive financial undertaking. Legislation from successive governments has failed to address the financial causes of New Zealand's – and the world's – housing crisis due to the significant (and increasing) amount of power the "FIRE Economy" – Finance, Insurance, Real Estate (Kelsey, 2015) has over Auckland's urban economy.

Chapter 5, marking the second half of the book, takes these conclusions about Auckland's relationship between unstoppable sprawl and a financialized urban economy to a national and then an international context. I note how financialized urban development is proceeding differently in Auckland's other suburban frontiers when compared to Kumeu-Huapai and show that the unending demand for "cheap" land to invest in is causing Auckland's urban economy to become a relevant force in people's lives in Aotearoa's most remote communities. From here, I present the two international case studies: lifestyle development in Melbourne's northern fringe running into economic and ecological barriers to further profitable expansion, and the corruption and inequality necessary to sustain unusually profitable suburban development in the California Central Valley. Drawing on existing literature for both of these examples, I use these cases to further underline the key point drawn from my study of Kumeu-Huapai: maximally profitable suburban development exists in proportion to its negative effects on human life and is one of our most immediate aggravating factors for rising urban inequality and failure to adapt to a changing climate. Chapter 6 brings this international view to the class dynamics of the urban periphery, pointing out the important role played by what Ehrenreich and Ehrenreich called the "professional-managerial class" (1977) in sustaining demand for large suburban properties and instituting particular forms of social separation to protect property values in new suburban areas. Yet even within this group of property owners, significant differences in social inclinations and practices between long-time rural-identifying professionals and newer residents (more cosmopolitan in their attitudes towards race and less loyal to local matters) create a dramatic rift in homeowners' visions of the town's future.

The last two main chapters deal explicitly with the pair of issues most drastically affected by the business of peri-urban development, beyond the reinforcement of class divisions: the collapse of local food production and vulnerability to climate change. Chapter 7 presents the issues of food security and food sovereignty. As people starve in New Zealand even as we produce ten times the food the nation's people require, I point out the ways in which the monopolization and concentration of agricultural production have been important factors in the collapse of local horticulture, another example of the "creative destruction" necessary for financial success. In response to this problem, I turn to the work of Catherine Brinkley (2018) on High Rugosity Cities, a model of urban-rural interaction that acknowledges the unique advantages provided to small-scale horticulture close to urban centres. Should state policy ever seek to align with local food security, Brinkley's insights provide a useful starting point for imagining a more sustainable urban centre. Finally, Chapter 8 addresses Kumeu-Huapai's response to a massive flood that struck the town halfway through my fieldwork, at the end of August 2021, and argues that colonial peri-urban development patterns the world over have exposed new suburban residents to significant risks from fire and flood in the search for cheap land to build on. This risk is becoming far more marked, as climate conditions begin to destabilize and extreme weather events become more common. Efforts to confront and ameliorate the issues of both rising inequality and vulnerability to a far less amicable climate must begin with a reassessment of how we conceptualize and use land at the edge of the city.

Beginning with the day-to-day work of peri-urban development today and memories of the collapse of horticulture in Kumeu-Huapai, this book moves from historical perspectives and political economic critique of suburbanization in northwest Auckland to international perspectives on the costs of financialized sprawl. By doing this, I seek to demonstrate how the very profitable work of transforming the urban periphery into suburban and lifestyle living opportunities for well-paid professionals is creating a series of urgent challenges for humanity and the planet and to point to possible alternative approaches to urban development in these places. It is here, where the city's financial economy seeks new opportunities at its frontier, where the future of urban life will be decided.

## Notes

1 At the periphery of urban areas.

2 Literally "People of the Land", Tangata Whenua is often used to refer to iwi or hapu (terms for groupings of peoples and families with connections to a particular geographical location) who hold a connection with the land being discussed, whether this be a particular location such as the Kumeu River Valley, or broadly to the entirety of Aotearoa. Two iwi hold Mana Whenua (lit. Power over Land) in Kumeu-Huapai today: Ngāti Whātua o Kaipara, descendants of the Ngāti Whātua who were expropriated of their lands in what is today the northwest reaches of Auckland, and Te Kawerau ā Maki, the survivors of the Kawerau peoples who had been pushed back to the Waitakere ranges by Ngāti Whātua centuries before.

## Bibliography

Aitken, R. (2020). "A machine for living": The cultural economy of financial subjectivity. In P. Mader, D. Mertens, & N.A.J. van der Zwan (Eds.), *The Routledge international handbook of financialization* (pp. 369–379). London: Routledge.

Auckland Council. (2018). *Auckland Plan 2050 Evidence report: Demographic trends for Auckland: Data sources and findings June 2018*. Auckland Council. https://www.aucklandcouncil.govt.nz/plans-projects-policies-reports-bylaws/our-plans-strategies/auckland-plan/about-the-auckland-plan/Evidence%20reports%20documents/evidence-report-demographics.pdf

Bourdieu, P. (1998). The essence of neoliberalism. *Le Monde diplomatique, December 1998*. https://mondediplo.com/1998/12/08bourdieu

Bourdieu, P. (2005). *The social structures of the economy*. Cambridge: Polity.

Brinkley, C. (2018). High rugosity cities: The geographic, economic and regulatory pathology of America's most non-concentric urban areas. *Land Use Policy*, *73*, 215–224.

Curran-Cournane, F., Carrick, S., Barnes, M. G., Ausseil, A.-G., Drewry, J. J., Bain, I. A., Golubiewski, N. E., Jones, H. S., Barringer, J., & Morell, L. (2021). Cumulative effects of fragmentation and development on highly productive land in New Zealand. *New Zealand Journal of Agricultural Research*, *66*(1), 1–24.

d'Amour, C. B., Reitsma, F., Baiocchi, G., Barthel, S., Güneralp, B., Erb, K.-H., Haberl, H., Creutzig, F., & Seto, K. C. (2017). Future urban land expansion and implications for global croplands. *PNAS*, *114*(34), 8939–8944.

Deloitte. (2019). *New Zealand's food story: The Pukekohe Hub*. https://www2.deloitte.com/nz/en/pages/primary/articles/case-study-pukekohe-hub.html

Fasianos, A., Guevara, D., & Pierros, C. (2018). Have we been here before? Phases of financialization within the twentieth century in the US. *Review of Keynesian Economics*, *6*(1), 34–61.

Goda, T., Onaran, Ö., & Stockhammer, E. (2017). Income inequality and wealth concentration in the recent crisis. *Development and Change*, *48*(1), 3–27.

Graeber, D. (2009). Neoliberalism, or the bureaucratization of the world. In H. Gusterson & C. Besteman (Eds.), *The insecure American: How we got here and what we should do about it* (pp. 79–96). Oakland: University of California Press.

Gudeman, S. (2016). *Anthropology and economy*. Cambridge: Cambridge University Press.

Harvey, D. (2007). Neoliberalism as creative destruction. *The Annals of the American Academy of Political and Social Science*, *610*, 22–44.

Heley, J. (2011). On the potential of being a village boy: An argument for local rural ethnography. *Sociologia Ruralis*, *51*(3), 219–237. https://doi.org/10.1111/j.1467-9523.2011.00534.x

Hicks, D., & Curran-Cournane, F. (2017). *Matching farm production data to land use capability for Auckland – Technical report 2017/020*. Auckland Council Research and Evaluation Unit, RIMU. https://www.knowledgeauckland.org.nz/publications/matching-farm-production-data-to-land-use-capability-for-auckland

Joint Centre for Housing Studies of Harvard University. (2021). *Home Prices Are Skyrocketing in Most Markets*. Retrieved September 13 from https://www.jchs.harvard.edu/son-2021-home-prices

Kelsey, J. (2015). *The FIRE economy: New Zealand's reckoning*. Wellington: Bridget Williams Books.

Mader, P., Mertens, D., & Zwan, N.A.J. van der (2020). Financialization: An introduction. In P. Mader, D. Mertens, & N.A.J. van der Zwan (Eds.), *The Routledge international handbook of financialization* (pp. 1–16). London: Routledge.

Oneroof. (2020). *225,000 NZ Homes Earned More than Average Kiwi*. https://www.oneroof.co.nz/news/225000-nz-homes-earned-more-than-the-average-kiwi-37599

Radio New Zealand. (2019, May 2). *Building on Prime Growing Land – Nine to Noon*. https://www.rnz.co.nz/audio/player?audio_id=2018693232

Richardson, B. F. (2021). The price we pay for land: The political economy of Pukekohe's development. *Journal of New Zealand & Pacific Studies*, *9*(1), 7–23.

Stats NZ. (2022). *NZ. Stat*. Retrieved May 18 from https://nzdotstat.stats.govt.nz/wbos/index.aspx

Waldron, R. (2019). Financialization, urban governance and the planning system: Utilizing 'development viability' as a policy narrative for the liberalization of Ireland's post-crash planning system. *International Journal of Urban and Regional Research*, *43*(4), 685–704.

# Part One

# Financializing the Kumeu River Valley

# 1 The rise of property development, the collapse of horticulture, and the restoration of financial power in Kumeu-Huapai

*Figure 1.1* A suburb in the making at Auckland's western frontier. Photo taken by the author.

"All I do is put out fires". At a cluttered desk in a field office, samples of paint schemes and veneers pinned to the walls of an otherwise empty show home for the benefit of prospective customers, a project manager for a housebuilding company attempts to explain the state of his craft. His description is hindered – and yet also reinforced – by the fact that he cannot speak to me for more than three minutes without fielding a desperate and harried call from a contact on a building site. Whether he is detailing the step-by-step process of delivering a suburban house to a mortgagee as we sit around his desk or looking back on his early days as a project manager while he drives me through one of the dozens of new developments under

DOI: 10.4324/9781003400325-3

way northwest of Auckland City, he is cut off mid-sentence. Each conversation follows a common structure: He greets the caller by their first name in a calm and friendly tone, listens to the problem they present, and assures them that the issue – missing materials, indisposed labourers, cancelled appointments, and supply chain failures – will be solved immediately. He then flicks through his contacts; calls someone up, greeting them just as familiarly as he had the first caller; and arranges someone or something to be sent somewhere. This task done, he calls up the original supplicant, lets them know their crisis has been averted, and quickly wishes them well and hangs up as they thank him for his work. While a folder of customer information sits on his desk between us, he finds a replacement worker for a job site, offering the substitute a $1000 NZD (New Zealand dollar) bonus for taking the trouble to travel out from the city. As he points to a recently poured foundation slab, he arranges a new timetable on a delayed shipment of wooden framing with the trade desk of a hardware store, deciding to have it dropped off on a Monday rather than a Friday to reduce the risk of theft if left unattended over the weekend.

This stream of requests, concerns, and logistical challenges never abates for the hour and a half I am speaking to him, and indeed, work has never been busier, even as coronavirus continues to disrupt work and life in New Zealand, "While we were locked down, we signed up another 17 [*houses*]". Each prospective home-owner is managed just as closely by the project manager as any building site, and he sets aside days of the week to make site visits and talk customers through one of the most stressful processes of their lives. This position as mediator between nearly every party involved in the construction and sale of a suburban house is unsurprisingly very taxing. "I'm six years in, I'm probably the longest serving project manager [*in the company*], they do burn out … when I first started, they gave me five [*houses*] and I nearly fell apart, now, now…". Earlier, sitting in his office, he attempted to explain his workday by pointing to a cartoon that has been set as the wallpaper of his tablet. A figure stands in the noonday sun holding a nail to their head. Above them, a massive hammer is frozen in a rapidly melting block of ice, and this hammer is tied to a rope that will surely pull it in an arc towards the character's head the moment the ice gives way under the heat of the day. Like many other professionals in the construction and real estate development industry I spoke to over the course of this fieldwork, he had once worked in housebuilding in an independent capacity. "From start to finish", he developed plots of land and built houses for clients, and this experience serves him well as a project manager for a nationwide housebuilding franchise today. Also, like many other professionals I interviewed, he asked me if I intended to use my research to get work in the property development industry. When out driving with him to survey a site, he reminded me that construction was booming, and demand has never been higher. The owner of another housebuilding company later asked me over the phone (unprompted) "Do you have experience? If you do it's very easy to get a job right now, everyone involved is very short [*of staff*]".

Like so many other cities across the world, Auckland is expanding rapidly. An urbanizing population (d'Amour et al., 2017) and a "wall of money" from investors leaning on real estate as one of the few dependable forms of profitable return

(Wetzstein, 2021, p. 1) are both leading the demand for more rural land at the periphery of cities to be developed for residential purposes. As with so many other growing urban centres, Auckland's economic activity has become ever more entwined with the act of land development (Kelsey, 2015, p. 67). House mortgages dominate the national credit limit (Centrix, 2021), and in the formerly rural towns like Kumeu-Huapai that lie at the periphery of urban sprawl, construction and real estate have long overtaken agriculture, horticulture, and forestry as the largest employers after services (Auckland Council, 2019, p. 13). After a historic low in the rate of monthly building consents issued in 2010 after the Global Financial Crisis, the rate of consents has tripled over the ensuing decade to overtake the highs of the mid-2000s (Stats NZ, 2021). To express this phenomenon with a concrete example, a sales manager of a construction company specializing in large suburban homes explained he had been with the company for more than 14 years and seen 350 houses built; 250 of those had been built in the past *four* years. The population of Kumeu-Huapai doubled between the 2013 and 2018 Census (Stats NZ, 2020), and in fact, between 2014 and 2019, Kumeu-Huapai saw the highest increase in property values in the entirety of New Zealand (Oneroof, 2020). Perhaps to head off a discussion of gentrification, a local real estate agent was quick to assure me that such a rapid jump in prices was an aberration, mostly driven by speculative interest "it's because this is very much an area in development, pricing is more volatile and speculative", but did note the area, especially locations of which they say formerly prospective buyers "wouldn't want to live there", had become "the primo suburb".

When looking at work and life in peri-urban Auckland today, one of the strongest initial impressions that will greet any observer is the massive transformative power over land possessed by development interests, their financiers, and other attendant professions. To begin the process of examining the phenomenon of urban development in Kumeu-Huapai, this first chapter looks directly at the rising star of the business of property development, in direct comparison and contrast with the former dominant industry in Kumeu-Huapai: horticulture. The unending demand for new land to be released for development across Auckland encourages the subdivision of land and makes less lucrative land uses economically non-viable long before the urban built environment is actually physically created (Curran-Cournane et al., 2021).

Thus, many people, myself included, can find it difficult to imagine that exurban development areas like Kumeu-Huapai were once prominent sites of peri-urban horticulture. One drives past an empty field, a tumbledown shed, a peeling wooden letterbox, and a faded "no trespassing sign", and if curiosity drives you to look up who exactly owns this patch of land at the edge of town, you are greeted with a single name, or an innocuously named investment company as the deed holder – Blue Sky, Golden, [*family name*] Trustees, [*full name*] Limited. It can be easy then, even in a colonial nation where multiple waves of expropriation and accumulation have occurred in the space of the last two centuries and in living memory, to simply imagine that this field was always here, empty save for summer-dried grass and weeds that must be mowed, waiting at the edge of this town for someone to level and compact the earth, install piping for water and sewerage, cables for electricity and internet access, pave a road, pour foundations, and build several large single-family homes.

Property development cannot be ignored, not only in the actual creation of new suburbs and the arrival of their new residents but also in advertising billboards, banners for open homes, traffic jams, and impermeable surfaces over the soil. Horticulture in Kumeu-Huapai, meanwhile, is most notable in its absence from the landscape save for a couple of remaining operations and some splintered greenhouses in fallow fields. The relationships between ideologies and the creation of landscapes have been a crucial aspect of critical theory since before Marx and Engels used Feuerbach's presentation of the cherry tree as an example of the ahistorical and self-justifying philosophy they sought to argue against (Marx & Engels, 1976, p. 45). Yet, before I discuss Kumeu-Huapai as a case study for the political economy of Auckland's urban development and of the axiology of peri-urban land, I will first present the changing face of life and work in this town through the eyes of these two industries and their relationship to the land and its people. To do this, we must understand what drives demand for housing at the urban periphery in the first place.

## Theorizing peri-urban development

When Pierre Bourdieu (2005) sought to dissect the social structures that defined the production and sale of large single-family homes in peri-urban France, he began with the simple position that "the social world is present in its entirety in every 'economic' action" (p. 3). Bourdieu held that to rely on a framework of analysis that paid no attention to how "the economy" was actually situated within social relations at every level – from production to circulation and consumption – would not only offer an incomplete picture of work and life but also actively occlude the driving forces behind the phenomena being observed, creating an "amnesia of genesis" (p. 5). His work on the French housing market, therefore, carried forward his own body of theory on the continuous and intergenerational accumulation of signification, associations, and responses that produce the sedimentation of behaviour and practice which he referred to as habitus. This set of inclinations will be very important for a discussion of class in the urban periphery in Chapter 6, but here it is most important as a way to understand just how much meaning is placed in a suburban house by its occupants. The home held a uniquely powerful position in the lives of Bourdieu's interlocutors, not only as an economic investment and as an attempt to lay the foundations for social reproduction (p. 21) but also as a repository for claims to desirable ways of living, jointly constructed by the people selling the house and those seeking to buy it (p. 24).

Bourdieu identified three key factors that defined housing supply and demand for the professional-managerial households intending to purchase a large, "authentic" suburban or exurban house while still commuting to the city for work. The first is the familiar situation that an urbanizing population and the financialization of housing had made purchasing a house in the city where the interlocutors worked very difficult, and they were thus forced to look further afield (p. 77). Secondly, state policy on housing and finance from the 1970s onwards had created a housing market specifically oriented towards housing as a private financial agreement between a bank, a developer and a mortgagee where home ownership hinged upon the latter party agreeing to become massively indebted to the former (p. 57). Lastly, sales and advertising agents for large housebuilding firms had seized on the aesthetic of rural "traditional" life as a strategy through which to convince urban professionals to settle for a house far away from where they worked and socialized, justifying long commutes, mediocre build quality, and high mortgage payments by "making a virtue of necessity" (p. 56). Sales representatives and advertisements appealed to a collection of symbols and practices (independence, ownership, and control over immediate surrounds) that are most relevant to precisely their customers: educated and well-paid white-collar urban professionals who will be able to keep up with mortgage payments but cannot afford to stay in the city.

The result of these three factors is the proliferation of the industrially produced "mason house" at the periphery of urban centres, and the condition that Bourdieu memorably calls "The Foundations of Petit-Bourgeois Suffering" (p. 185). These professionals, in seeking a lifestyle that matches their ambitions and meets their budget, are:

> lured by the mirage of falsely "individual" housing … experience neither the solidarity of the old working class districts, nor the isolation of the better-off areas: these people, who spend hours each day commuting to distant workplaces, are deprived of the relationships that formed within their neighbourhood … The single-family house thus functions as a trap … .
>
> (p. 189)

Bourdieu thus presents an analysis of the housing market in peri-urban France that does not simply provide theories as to drivers of supply and demand in aggregate but seeks to incorporate the motivations of financiers and developers, the formation of state legislation, the dreams and desires of the class of people who buy the houses Bourdieu is looking at, and the understandings that professionals in the real estate sector have of their target customer. Bourdieu's research comprises the first part of the theoretical basis of this chapter precisely because of this approach, and my own presentation of the ascension of property development and the collapse of horticulture seeks to incorporate aspects of his methodology.

The second half of the theoretical context that underpins my analysis of peri-urban development comes from Stephen Gudeman's (2016) model of economic activity as a series of concentric spheres and his resulting argument that even the most rarefied and abstract forms of economic activity depend upon the existence of the household and the community for their own existence and success. Drawing on his work in rural villages in Latin America and the "advanced" economies of international cities, he described economic activity as being composed of five spheres, from the unprofitable yet vital work of sustaining a household, to the sharing of labour and resources among a community, then to commercial activity, finance, and finally the rarefied realms of meta-finance. At each ascending level of economic complexity, the exchanges involved become ever more calculative and ever less mutualistic, yet financial profit-making can only occur if the household and the community, the foundations of the economy, can continue to provide its means of exploitation (p. 167). Such a contradiction, in Gudeman's view, leads directly to the crises of inequality, financialization, and household indebtedness that plague the global economic system today (p. 175).

This diagnosis is particularly useful for understanding the fraught relationship between a house as a place to live and a house as an ever-more-important financial asset class in New Zealand markets and across the planet (Fernandez & Aalbers, 2016). The result of this conflict is that suburban development is guided by the imperatives of the latter as much as the former. The conclusions that Gudeman draws from his model have precedent in geographical and historical work on global capitalist development, notably the existing corpus of Marxian theories of social reproduction under capitalism (Harvey, 2010, p. 149) and Giovanni Arrighi's world-historical depiction of global financial empires as entirely dependent on a productive base which they inevitably hollow out and in doing so destroy the basis of their success (Arrighi, 2010, p. 325). This model of interrelated spheres of economic activity provides a useful approach to an analysis of urban development in Kumeu-Huapai that complements Bourdieu's focus on the intersection of the visible material economy and the collection of learned inclinations and behaviours that delineate social class. Taken together, my approach seeks to capture the relationships between government, financiers, landowners, land developers, house-building companies, tradespeople, real estate agents and customers that are shaping the explosion of residential development in Kumeu-Huapai, and how these same relationships made peri-urban agriculture non-viable by the new millennium and are drastically changing work and life for everyone living in the area.

## The business of residential development

Seeking information on the sale of a 6–7 hectare kiwifruit orchard sold for 5.71M NZD, I asked the real estate agent who brokered the deal about what the future of the land would be:

> It'll be land bankers, they see that as potential growth. You can use [*rent*] the house, the shed, the orchard, make a 1.5 per cent return on something that otherwise doesn't cost you much at all. In five years' time it'll be worth ten million, in ten years' time, thirty million.

This same real estate agent was highly placed in local politics a decade ago when local governance was reconstituted into a single "supercity" and council and development in Kumeu-Huapai began to accelerate. They freely admit they were "lucky enough to make some sales at the same time" by combining their knowledge of the change in zoning regulations bought about the reorganization and their professional position in the Kumeu-Huapai real estate market. To underline the importance of insider knowledge, they recount a story of a local man who had recently sold his family farm for $4M NZD, and understandably very satisfied, left to retire, only to hear some months later that the buyer had an insight into future development plans and had subdivided the land to sell for $69M NZD.

Two initial facts about the business of residential development at Auckland's urban periphery that jump out at any observer are that a lot of money is being made by some parties and that these parties are successful partly by virtue of connections across several sectors of working life. A construction company owner boasts that they play golf with a director of Cabra, the largest land developer in the region, in response to a question about future land supply for building, while this director himself tells me about the small cluster of specialist firms Cabra has worked exclusively with for decades, and a local government official informs me with some amusement that this very same director called him on the phone to ask some innocent questions about the history and ownership of some farmland adjacent to the Kumeu-Huapai town centre. Throughout the course of my fieldwork, particularly when it came to speaking to professionals whose work in any aspect even tangentially related to residential development in peri-urban areas, information about development proceedings or legislation was rapidly transmitted through informal channels well before it ever made it to the media or a public announcement was made.

Often, in learning about recent events in the development sector and discussing them with interlocutors, I was part of the process of knowledge dissemination. More than one interviewee remarked they were happy to speak to me specifically because I was openly discussing what I had heard from other professionals, leading me to consider my words more carefully as fieldwork went on. On one occasion, while asking a disinterested sales representative the kind of obvious questions only an anthropologist can provide, he suddenly started to attention, looking at me over his vape held in front of his face like a Dictaphone, and asked me "Where'd you hear that?". My question included a reference to Fletcher Residential, the largest housebuilder in the country, having purchased rural-zoned horticulture growing land adjacent to the Riverhead town centre northwest of Kumeu-Huapai. The transaction had not been officially confirmed at the time, but its ramifications were already being discussed by local government and development professionals.

Access to this information is no less important for landowners seeking to make a fortune, as shown by the previous anecdote of a $4M NZD farm that became a $69M NZD development in a matter of months. A local government official presented their own story of one old couple as their vision of an archetypal successful land sale. "They'd always quietly turn up the ratepayer's meetings, and National Party meetings, and then once the SHA came through they turned up in a brand new merc, and they were in their 80s and he had new trainers on". The "SHA" mentioned in this quote is one of the dozens of "Special Housing Areas" created by the National government from 2013 to 2017, in which developers worked with landowners to petition the government for areas of land to be rapidly made available to develop and build on. One of the SHAs in Kumeu-Huapai, now referred to as the Huapai Triangle and developed by Cabra, was the first and largest in the country (see Figure 1.2). Real estate agents, planners' community action advocates, and landowners all told me that before it had been designated an SHA and made available for 1,200 homes, it had been designated to be industrial land. The area was legally transformed into an SHA through the co-operation of the half-dozen landowners of the site with developers who possessed the necessary contacts and relationships to present the case before the central government and override local and regional planning decisions and massively profit. “That’s something the National Party does very naturally”, one planner explains, going on to describe their experiences of close communication between land developers, road building firms, and other related companies all through the conduit of the National Party. In this way, landowners at the edge of Auckland’s urban expansion are indeed active agents in the future of urban development (Murphy, 2020), yet their most dramatic successes only come with the alignment of a variety of state and market interests in their favour, and therefore must be both informed and connected to make the greatest profit.

It is not surprising that such a dense network of agents should exist around the residential construction industry in New Zealand, given the complexities and capital involved in building an entirely new suburb at the rural periphery of a major city. Take the land that makes up the Huapai Triangle today. A more comprehensive history of the area will be presented in the following chapter, but for the purposes of the most recent landowners, this was once the site of the Nobilo’s winery. Founded in 1943 by a recent emigrant from Croatia, Nikola Nobilo, the vineyard

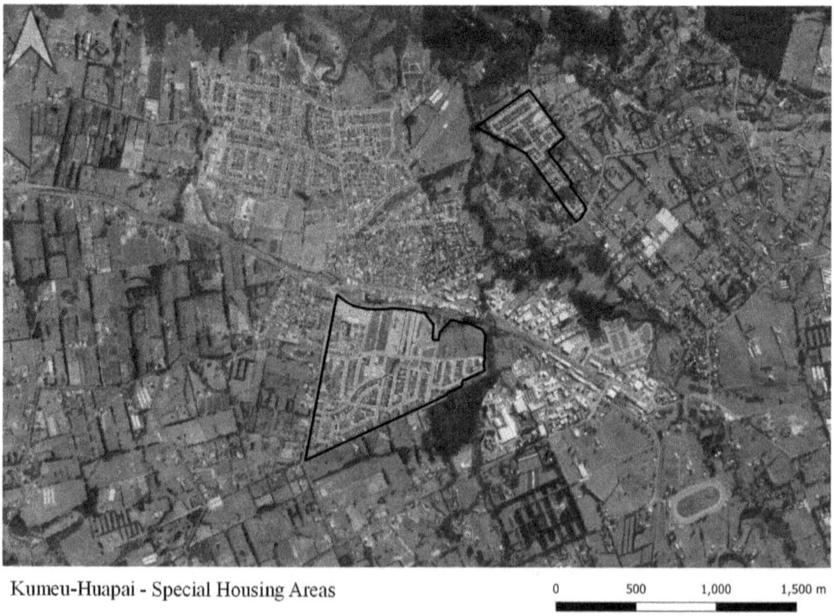


*Figure 1.2* The two Special Housing Areas in Kumeu-Huapai, the larger of which is the Huapai Triangle. Zoning data from Auckland Council (2016b) and satellite imagery from Google Maps (Google LLC, 2022).

was successful for decades and launched the Nobilo name. The label is now owned by the international brand conglomerate Constellation Brands, and Nobilo's Sauvignon Blanc is sold across the planet. But industry and family had made their marks on the operation, such that by the time of the Special Housing Areas, Nikola's son was one of half a dozen landowners of the former vineyard site now slated to become an industrial area. Seeking vastly greater profitability, they sought to make the area available for residential development instead, and this ambition is where the process of peri-urban development begins for most involved.

If any landowner at the edge of the city hopes to extract this massive payout from suburbanizing the land, they must first overcome the legal and logistical challenges inherent to such a transformation. Rural land must become zoned to allow it to be subdivided and for housing to be built on it, and any applicant – be they from local government or a private organization – must present reports from experts that attest to positive results and a mitigation of negative factors. Engineers, environmental scientists, and planning consultants are all brought in by the applicant to present their own expert opinion on the ways in which the zoning change follows national-level legislation such as the Resource Management Act and regional policies like Auckland's Unitary Plan, and that development would have a net positive effect on the area. This requirement to pay for expert opinion on a plan change necessarily restricts this process to local government and the large landowners and developers capable of committing the resources to such a process, and has led sole trader developers (like the project manager at the start of this chapter had been) to find work in larger companies. Should the change be approved, either at the level of Auckland Council or by central government overriding the process, the first visible act of urban development is the surveying and development of the land itself. An environmental lawyer places the land developers (and their attendant specialist companies) at the centre of the entire process due to their role at this point: "they [*Cabra etc.*] know how to do it, to turn bare land into developed land, and that's because they have a massive machine behind them". It is tangential, but perhaps relevant to understanding land development in Kumeu-Huapai, that a major shareholder of Cabra and brother of one of the original directors is also the director of Hiway Stabilizers, New Zealand's only major road builder.

Land developers seek finance from banks, often with the expectation of a minimum 20 per cent return (Murphy, 2019, p. 7), and once this is gained, they put their considerable resources to work. Earth is moved, pipes laid, roads paved, and streetlights fitted (see Figure 1.3), and by this, labour rural land is transformed into a suburb-in-waiting, wanting only for the actual houses to be constructed. Now comes the time of the housebuilders, the real estate agents, and the clients. Preferably, the client agrees to buy the section of land and pay for the construction of the house by the housebuilder together as package, generally getting a mortgage from a bank to do this. This ends the involvement of the land developer, who, according to multiple interlocutors, normally sell the land for approximately double what they bought it for. Before the customer can walk into their new home, however, a wide variety of individual tradespeople of various skill sets must be independently hired and supplied with materials. A checklist given to me by a project manager lists 206 separate tasks (see Appendix 1 for the complete list) to be done, from surveying and earthworks all the way to personally handing over the keys to the homeowner. During this building process, the site is checked by local officials multiple times; temporary toilets, water tanks, and rubbish skips are set up; and a project manager individually arranges a task – "Foundation Prep", "Frame pre nail", "safety nets" and 203 more – to be done by a certain contracted and scheduled team. When this process ends, there stands a suburban house on a paved street where a few years previously lay rural land. That peri-urban development in Auckland would simultaneously involve panels of academic experts, international banks, corporations with a monopoly of scale over urban infrastructure, and hundreds of independent contractors and suppliers drawn from across the region and kept on mostly on the basis of personal understandings between the contractor and the manager is testament to the close working relationships across sectors that make such a multi-faceted enterprise possible.

A critical aspect of fieldwork that will define this chapter and this entire book is that almost all *formal* interviews were undertaken with interlocutors in the professional-managerial hierarchy of real estate development companies rather than their contract workers. Following ethical advice to approach organizations from the top down and seek permission to speak with employees from their managers (The University of Auckland, 2020, p. 57), I found opportunities to speak to workers on the occasional building site but did not receive access to contact staff

*Figure 1.3* Land being graded, and roads, pipes, and streetlights being installed in the Huapai Triangle by workers contracted to Cabra. Photo taken by the author.

formally. Often, I would be handed a contact for a supervisor who would not reply, but I was also directly told by some managers that they either did not themselves deal with the relationship between the company and its labourers, or I was simply told "I will not grant you access to my team" in response to such questions, ostensibly because I would be wasting their time. This lack of formal opportunities for conversation with construction workers and their supervisors makes this project very much a study of property development professionals, rather than of the people who supply the actual labour that enables these professionals to succeed.

From the conversations I did have, workers explained they are contracted on a mostly exclusive basis to a major housebuilding firm and work across the region for them on a schedule set by their project manager at the company. When the project manager at the start of this chapter called up someone they knew and told them they would give a tradesperson a $1,000 NZD bonus for travelling out of the exurban site at short notice, they were speaking to a supervisor who would in turn let the workers under their purview know about the offer. I was unsurprisingly also told that people at the supervisory and managerial level had very little incentive to present the day-to-day realities of the work to me honestly, that I should be aware that niceties such as Occupational Health and Safety and workplace COVID precautions necessarily had to be bent or outright circumvented to keep the process moving, "they don't want to really know what we do to make it work". One story I heard that illustrates the necessarily improvisational and non-regulated nature of the work was that (to be as vague as possible) a housebuilder in the Auckland region has bypassed the requirement that a construction site be inspected by local authorities by claiming to any who asked that they had a waiver over the need to have their foundations checked, though no such waiver has ever existed.

Today, the relationship between employee and employer is troubled for more than usual reasons; a report on the construction sector in 2021 noted that "we are heavily reliant on immigration to fill our needs" (BDO, 2022, p. 13) and that employers are demanding the end of COVID-19 travel restrictions to bypass a local workforce that has begun to take advantage of the relative scarcity of labour by leaving for better-paid jobs mid-project. The number of people in New Zealand working in the construction industry has increased by 8 per cent in the past year (Ministry of Business Innovation and Employment, 2022), workplace injuries on building sites increasing and becoming markedly more costly over that same year (ACC, 2022), yet this work is compartmentalized and anonymized in the work of a manager who thinks in terms of half a dozen ongoing projects at any one time and can only make brief site visits on a weekly basis.

Just as much of the labour involved in residential development can be occluded, so too can the nature of the building materials. Everything from custom-built wooden frames to water mixers for shower units are purchased through the trade desks of a few large franchises that in turn arrange the import of these goods through a number of import/export outfits of varying scale and specialization, who *in turn* are seeking to purchase these products from manufacturers, a markup for profit being added at every step of this process. Like much else in working life, this complex system has suffered repeated upsets and holdups over the course of the pandemic as factories in China and Southeast Asia were forced to slow production and export of materials to nations like New Zealand. This country imports 90 per cent of the materials that go into a house (EBOSS, 2021, p. 6), and the result is cost increases "in double digits" percentages (BDO, 2022, p. 3), a general increase in lead times for all shipped goods and delays for crucial materials like laminated boards longer than 12 months, according to one project manager. The other half of this equation is that demand even among existing clients of suppliers (the aforementioned nationwide trade supply chains) has only continued to increase in New Zealand and across the world (EBOSS, 2021, p. 12), as the pace of global urban development accelerates beyond the pace at which factories and ports can operate. A house built in Kumeu-Huapai today is thus (materially at least) a product of global structures of production and logistics as much as it is defined by regional labour and the goals of national-level banks, land developers, and housebuilders.

## Whose business?

I am sitting in another show home, in another new suburb, speaking this time to a sales manager of a housebuilding company. Sitting on the couches in the two-storey home's front room, he takes a call before we start the interview. The caller is a prospective buyer and is interested in speaking to the sales manager about the process of working with the design team to have a home of their preference built on land in Kumeu-Huapai, until the sales manager informs them the last house and land package he sold of a similar type out here went for 1.45M, at which the caller makes an excuse and hangs up. "Some people think you can still buy a section out here for 1.1" they remark to me. While the domination of suburban patterns of development over denser urban forms in Auckland is partially driven by the greater costs and complexities of doing the latter (Ministry for the Environment and Stats NZ, 2021, p. 8), this supply must be at least vaguely connected with a corresponding demand, so who are the people who move to these new suburbs out here at the edge of the growing city? Real estate agents and sales managers for housebuilding companies described a rough 70-20-10 split between those who already owned a home and were selling it to move to Kumeu-Huapai, those who purchased a house entirely as a financial product, and a remainder of first home buyers with an unusually high amount of income. This first and most numerous contingent are the people fortunate enough to have been homeowners during the massive appreciation in property values over the last two decades (Reserve Bank of New Zealand, 2021), now seeking to trade up for a bigger place to live:

> clients have mostly moved here from the north shore and the inner west … lots have come from Te Atatu and Henderson … a desire for more space and a nicer lifestyle … a lot of people come here from cross lease, apartments, or terraces.

Advertising efforts from developers, builders, and agents promote (and exaggerate) the relative proximity of Kumeu-Huapai to the city and the western coast as well as recreational amenities such as vineyards and forests (see Figure 1.4) to capture specifically this clientele.

The second main group, however, the investors, are attracted to the particular type of new housing found in peri-urban developments like Kumeu-Huapai for strictly non-aesthetic reasons. "There's two types of buyer, an investor wouldn't pay half the attention [*to the style of the house*]", prospective investors are instead seeking houses that are brand new and thus will have low maintenance costs and are in areas subject to rapid appreciation and population growth. These two groups of buyers do not necessarily have mutually exclusive aims, however. One prospective homebuyer at an open home event explained that they had moved into a brand-new home in the Huapai Triangle two years ago, and was now using the appreciation of its value to sell up and move several blocks over into a yet larger house. The final and smallest group are the first homebuyers who have enough cash and/or regular income to meet a bank's requirement to finance nearly 1.5 million dollars, "[*they*] need a lot of clout … 130k in cash" one real estate agent conceded when asked about the affordability of the new housing. For these reasons, most of the new residents of Kumeu-Huapai are salaried professional-managerial workers that drive into the city to work, having moved in from elsewhere in Auckland (Stats NZ, 2020). This final aspect of the business of property development in Kumeu-Huapai, the customer, has as much of an impact on work and life in the area as any inclination or decision made by a bank, landowner, developer, or housebuilder.

With the development process of a new suburban house built on formerly rural land in Kumeu-Huapai laid out in this way – from the possession of the land by someone intending to make the maximum possible profit, to the legal and juridical process of allowing and enabling development, to the actual act of land development, the doling out of parcels of the land to clients to be worked on by housebuilders, and finally the finished house being occupied or otherwise used as an asset – how does it relate to work and life here? As with my description of development, my theoretical discussion of its influences on social and productive life begins with the state of the land itself. Just as a local government official told me they had been receiving calls from a director of Cabra asking after the provenance of some farmland near Kumeu-Huapai, those who buy and sell land must be keenly aware of state and developer visions for future growth, and land use strategies for peri-urban landowners tend to become increasingly oriented towards a future hypothetical subdivision and sale for residential development. Real estate agents selling rural "lifestyle" blocks are sure to advertise the possibility of urban expansion and the concomitant massive profit-making (see Figure 1.5) as much as they stress the seclusion and bucolic aesthetic of the site as it exists today, and "land bankers" look to the overgrown orchards and the empty livestock pastures surrounding Kumeu-Huapai's town centre as a crucial investment avenue.

Just as the real estate agent/former local politician at the start of this section described what would happen to the kiwifruit orchard they had sold, residents are noticing the gradual replacement of resident landowners with absentees who rent out the purchased property until such a time as it can be sold to a land developer. Information about lucrative blocks being up for sale is communicated through informal

*Figure 1.4* Advertisements for a development in Kumeu-Huapai emphasizing local amenities and the close proximity of Muriwai Beach and the Auckland CBD (Kumeu Central Property Limited, 2022).

channels in the same way as any other part of the development process, and one local government official, due to their own connections to the Chinese community in Auckland, told me that Chinese-language newspapers often feature advertisements offering to act as agents in land banking process, resulting in a concentration of non-European land bankers in some parts of an otherwise overwhelmingly Anglo exurban area (see Figure 1.6), a long-time real estate agent echoed this observation; "99 per cent of them [*the new residents*] are kiwis [*of European descent*], but of course it was Chinese people who were banking all the land in the first place".

In this way, the peri-urban land of Kumeu-Huapai becomes a financial product into which individuals and communities place their capital in motion in hopes of a huge return on investment.

The course of capital accumulation never did run smooth, however, and this is most especially true when it comes to investments that take many years or even decades to come to fruition (Harvey, 2010, p. 42). I was told by the same local government official who had been reading Chinese-language papers about land banking that they had attended an emergency meeting in the north of Auckland when a large number of Korean land bankers, all of whom had bought land on the promise from the agent that it would become suburban housing in due time, had woken up one day to discover their investment was instead to become industrial

*Figure 1.5* Real estate agents advertising the future development possibilities of rural "lifestyle" properties. Photos taken by the author.

and commercial land. "I've never met so many Koreans turn up to a meeting to lose their shit!" they exclaimed, telling me that, just as with the Huapai Triangle, some of the major landowners would doubtless be seeking out a developer to aid them on an application to have this zoning decision changed to one that better suited their interests. The value of peri-urban land to developers and their financiers as raw material for development thus encourages its utility as a financial asset in growing cities and has a powerful effect over its current and future uses.

Even as Auckland's urban expansion has made Kumeu-Huapai's land more valued for financial ambitions than productive purposes, it has reoriented social life in the area ever more towards the city. The most visible cause and expression of this realignment is of course the number of Auckland residents deciding to move further out in search of a larger property and a rural lifestyle, while maintaining proximity to their work and social lives in the city and inner suburbs of Auckland. Yet, this is accompanied by a corresponding lack of local employment opportunities and public transport options – both far less profitable to create than suburban residential development – that serve to render the new frontiers of the city entirely reliant on its core. One community advocate simply told me "We didn't want to be

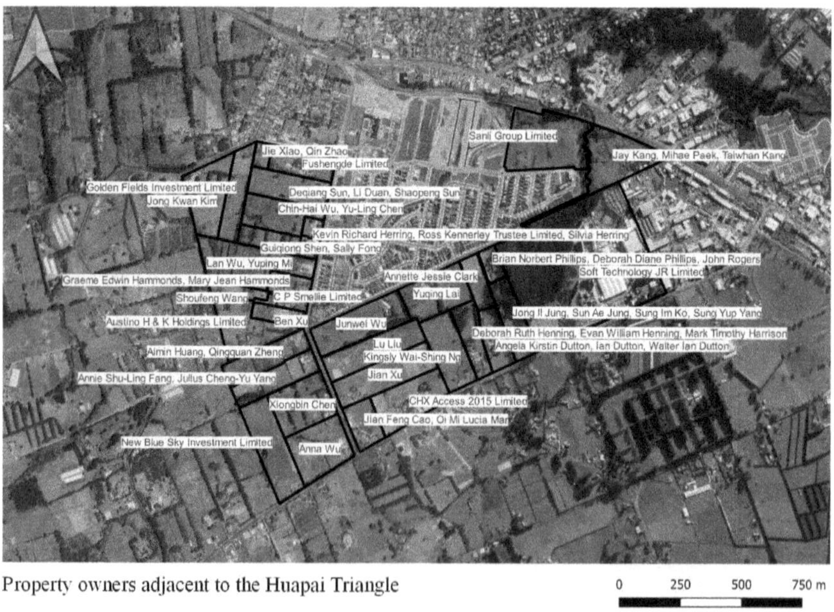


*Figure 1.6* Landowners immediately adjacent to the Huapai Triangle, one of the land banking areas pointed out to me for having been shared among Auckland's Chinese community. Note that this is a localized phenomenon and that nearly every other aspect of the development process is owned by European New Zealanders. Chinese land developers and housebuilders do operate in Kumeu-Huapai but have a small share of the market. Property ownership data received from Land Ownership New Zealand on request (LINZ Data Service, 2020). Satellite imagery from Google Maps (Google LLC, 2022).

a dormitory suburb of Auckland, but that's what happened". A volunteer firefighter commented on the slow change among fellow volunteers from mostly working in local primary industries or technical trades to almost entirely being employed in professional managerial roles and commuting into urban Auckland ever day; "when I joined, most all the people in the brigade were in trades … now there's one guy in trades and everyone else is white-collar workers". The transition from community life centred around local town halls in the middle of the 20th century was a slow process, aided as much by the dominance of the automobile from the 1970s onwards as it was by employment and demographics (Dunsford, 2002, p. 238), yet a brief period, often pinpointed by locals specifically as the last seven years, saw a rapid change in day-to-day life. As interest in exurban lifestyle block living continued to increase, a Cabra development on Tapu road, land banked in the early 2000s and begun around 2010 (see Figure 1.7) began to sell newly built houses and the two SHAs saw their first foundations laid. This led to the doubling of the population and unparalleled nationwide jump in land and house prices that defined real estate property relations in Kumeu-Huapai in the late 2010s, as the promise of future development created financial demand for land, and real estate professionals continued to sell the promise of exurban living to those who could afford it.

In a relatively short time, daily commute traffic became the single greatest popular issue faced by local representatives, the MP for Kaipara ki Mahurangi telling me "nine out of ten people that come into this office are here to complain about the state of roads" and estimates "from my own observations at the side of the road" that the average car driving through Kumeu-Huapai "has 1.1 occupants". On another occasion, one resident claimed they had turned to listening to motivational self-help CDs in the car on the way to work to remain calm. High levels of traffic and very visible urban development have become a negative factor in selling the area to prospective buyers from the city, one agent complaining "people come around on a Sunday to look at an open home and then get stuck in traffic on the way back". Such a self-inflicted wound on the property sector's presentation of Kumeu-Huapai as "lifestyle block paradise" led to many real estate agents, developers, landowners, and other interested parties demanding the accelerated funding of public infrastructure projects in the area and informing me (often in defence of their profession) that development in the area would surely slow down, even as others spoke of massive organizations like Fletcher acquiring rural land nearby and intending to use their relationship with the central government to gain permission to develop it. The success of removing barriers to peri-urban development and then selling an exurban lifestyle to people moving out from the urban core has in turn irrevocably changed Kumeu-Huapai into a node of Auckland City, materially, socially, and financially.

As real estate agents explained how they sold houses in Kumeu-Huapai and the surrounding area, they would often acknowledge that a new paradigm was eclipsing the story that had very recently been used to sell peri-urban living. "The quarter-acre dream is well and truly over" one land expert told me, before saying that an important aspect of new builds would be a house (single family or not) that covered much of the plot of land it was situated in, as activities like gardening

*Figure 1.7* Part of Kumeu-Huapai in 2010 and in 2016. Note the major development (Tapu road) at the centre and left of the image, and the cluster of new lifestyle block houses on the right. Satellite imagery sourced from Auckland Council's GeoMaps Service (Auckland Council, 2016a).

would be a luxury of time new owners were unlikely to be able to afford. It was repeatedly brought to my attention by many interviewees that childcare centres were being built in peri-urban Auckland just as fast as new subdivisions were being announced because most of the professional-managerial class families who moved out here consisted of two parents who both worked at least 40 hours a week and their very young children, and they did not have nearby extended family to rely on for help raising these children. Just as the profit-maximizing strategies of land developers and financiers in the Auckland region have a powerful effect on the resulting shape of urban development for the rapidly growing city, so too has the intersection of the capital available to professional-managerial homeowners looking to find a better lifestyle and the motivations of real estate agents seeking to sell them on places like Kumeu-Huapai created the particular conditions of life for new homeowners here. The process of sprawl and suburbanization that absorbs formerly rural towns into the greater metropolitan system, the financialization of land and housing at the urban periphery, and the gentrification of exurban areas all possess a deep relationship to the business of residential development itself. As the following section will demonstrate, the key to this dramatic social and economic transformation is the disappearance of the rural order that had once dominated.

## The collapse of peri-urban horticulture

It is the 99th annual Kumeu Show, and the showgrounds of the Kumeu Agricultural and Horticultural Society play host to a similar collection of attractions as can be seen at any Agricultural and Pastoral Show in Aotearoa: spaces in the wide fields are marked out and rented to stalls selling food and trinkets, the Show Committee has paid for a number of fairground attractions to be set up, and a series of rural-themed competitions – woodcutting, livestock presentation, and sheep shearing – form the aesthetic centrepiece of the day (see Figures 1.8 and 1.9). Limitations and cancellations caused by the pandemic led the organizers to shorten the show from a weekend to this single day, as it once was decades ago. The decision to curtail the show irks the stall vendors the most, one telling me that the two-day setup allows parents to bring their children along on the day when they aren't ferrying them off to sports or social activities and that these families are the vendors best customers, the reason they set up shop here every year in the first place.

Certainly, families with young children are the majority of attendees, and these stalls are as much a centre of attention as the sight of a prancing horse or a practiced lumberjack. Yet, some stall owners – in particular, those presenting glossy cars, freshly painted tractors, and other items of heavy equipment – feel that their time here is wasted. "The Kumeu Show used to be a place to make deals" a car salesman mutters, as passers-by glance at the 50,000-dollar four-wheel drives parked in a neat line beneath banners bearing the brands available from the dealership. He goes on to tell me that he mostly shows up here every year to continue the tradition of attending rather than for any commercial motive. An agent for a farming equipment importer enthusiastically informs me from behind a camping table laden with chainsaws that "this used to be the biggest one-day show in the southern hemisphere", a leatherworker remarks somewhat bitterly that the sheer popularity of the show in decades past required stall owners to book their space for next show immediately after the current one had finished and that now there are empty spaces visible. Only a single stall sells any produce, and the seed drills, excavators, and scissor lifts mostly serve as interesting objects for children to climb on.

The way one pair – an older man and a younger woman selling belts and purses – describes this show, it peaked as an agricultural and horticultural event in the 1980s, collapsed over the 1990s, and was reborn as more explicitly fair-like attraction, with ever more space given over to market stalls and ever less to the services that sold to productive agricultural operations. As we stand in a crowd watching teams of shearers from across Aotearoa deftly drag, flip, and shear stunned-looking sheep at a rate of less than a minute each, one of the main organizers of the show agrees

*Figure 1.8* Wood chopping and shearing competitions at the 99th annual Kumeu Show. Photos taken by the author.

*Figure 1.9* Clothing and cars for sale at the Kumeu Show. Photos taken by the author.

with the stall owners' comments once I bring them up and, motioning to the ongoing display and the people clustered around it, tells me this offers the attendees "an opportunity to reconnect" with the forms of agrarian life that once defined places like Kumeu-Huapai. Even as this area becomes ever more defined by its relationship to the city centre and less by rural productive activity, events like the Kumeu Show are artefacts that provide an insight into the recent history of peri-urban life.

In my attempts to learn about the state of actually existing peri-urban horticulture in Kumeu-Huapai, a consistent challenge from interlocutors to any enquiry was that horticulture was entirely gone. "Drive up some driveways and find out, Huapai to Foster road, no one's picking apples around here … now the families are fighting over what to do with them" one real estate agent told me, claiming that the financial value of land in the area was such that when land bankers came in, growers "took the money and ran", and that the land bankers would "buy places for 2.5m, then sell for 4m a few years down the line". A local government official remarked that I shouldn't expect anything else from horticulturalists, saying that the work had normally been done by immigrant families with few prospects other than working the land and would immediately sell the land if it became financially lucrative to do so.

Like so many other older professionals in Kumeu-Huapai, both of these interviewees had previously worked in horticulture and were not speaking entirely without experience in such matters. Small-scale peri-urban horticulture – where areas of growing land of a half-dozen hectares or less at the city's edge could make take advantage of proximity to urban markets, labour, expertise, and materials – is invisible in Kumeu-Huapai (with a few exceptions) because it has been almost entirely eradicated, and its presence continued to decline further over my year of fieldwork. Asking at the Kumeu Agricultural and Horticultural Society about any members who still sold produce for a living, I was directed to one man who was attempting to stay commercially viable by renting the orchards of other collapsed enterprises and running them simultaneously with his own; "leased five orchards overall, one time it was four at a time". Yet, by the time we sat down in the meeting room of the Agricultural and Horticultural Society's office, he had sold his land to an investment company and decided to work in green waste disposal instead. The work had simply been too much, especially without other horticulturalist families nearby to control the spread of pests or parasites and lend help or equipment on a reciprocal basis when needed. "You need a community of apple growers" he explained, and as the only grower under the age of 70 left in the area desperately attempting to work multiple orchards, he saw no alternative to selling.

"It's all economics" this now-former horticulturalist told me on another meeting at his orchard, which he continues to mow and maintain after selling. Looking over rows of apple trees, still bearing fruit but now only feeding birds and rodents (see Figure 1.10), standing in a block of land now under the legal ownership of "Bangtai Investments", the simple formula of rising land prices bringing an end to the viability of horticulture in Kumeu-Huapai does ring true, yet it is far from the complete picture. Just as the accelerating pace of Auckland's urban sprawl is the result of a confluence of state and market interests acting on every level from local to international, the collapse of peri-urban horticulture was driven by the interests of individuals in horticulture, governance, property development, and finance who sought to make their fortunes as the consensus on economic policy began to shift across the planet.

The story, relayed to me from several former horticulturalists, a few land developers, and two local historians, goes like this. Horticultural production across New Zealand in the 1970s was managed by a series of centralized boards, the most relevant one in Kumeu-Huapai being the Apple and Pear Board. The Boards took fruit and vegetables from growers and paid them a stable price. Producers were not forced to compete in the market for sales, and the centralized character of a Board allowed them to provide small growers the economies of scale necessary to get access to heavy equipment, packing sheds and other useful forms of labour-saving investments and infrastructure without having to pay for it themselves, yet what they are most remembered for in the minds of some former growers was their enforcement of exclusivity. Attempts to bypass the Boards by selling produce under the table at higher prices were investigated and fined. Stories abound of inspectors chasing after truckloads of fruit down country roads, of orchards with multiple tonnes of fruit mysteriously unaccounted for, and of flourishing black market grocery operations for buyers in the city.

*Figure 1.10* Apples continue to grow and fall in an unpicked orchard less than a kilometre away from the Huapai Triangle. Photo taken by the author.

In hindsight, the Boards were the best deal the horticulturalists ever had in New Zealand. "A&PB was the best thing since sliced bread, it was fantastic" one former member recalled. Yet, many growers in Kumeu-Huapai and the surrounding northwest area wanted to make more money through deregulation. In this discontent, some larger buyers, growing operations, and financial agents saw an opportunity. When the neoliberal turn in politics saw the restoration of financial profitability and protection of market interests as the sole aim of the state (Harvey, 2007, p. 27), New Zealand was an early and enthusiastic adopter (Kelsey, 2015, p. 123). Growers suddenly found themselves having to compete with industrial-scale operations here and overseas, something they had been protected from for decades. Global shifts in supply and demand, or in the relative values of currencies, began to make themselves known in orchards in New Zealand. The Apple and Pear Board was forced to make its shares available to be purchased, allowing the corporate raider Ron Brierly and his firm GPG to buy out horticulturists struggling to survive in the new economy and assume control. Other public goods that had made peri-urban agriculture possible, such as subsidized rail freight for produce across the country, were removed as the priorities of the state changed.

The attempted privatization of New Zealand's fruit export business by the likes of Brierly did not, however, result in a successful seizure of profit streams, but in a generalized collapse in the former Board's production and circulation infrastructure. One former grower who sold their land long ago is still bitter about the privatization move "ruining the Apple and Pear Board to build an empire that never got realized". A particularly bad collapse in international demand in 1996 was the final straw for many fruit growers, and by the turn of the new millennium, most of the damage had been done. The orchardist who'd been renting other failed orchards tells me he'd been part of an attempt to form an Auckland Fruit Growers association "from Te Hana (north of Auckland) to Pukekohe (south)". It had 100 members in the 90s, but "we had 7 members by the end", and the group was effectively disbanded.

A Master's thesis in geography on peri-urban horticulture in northwest Auckland, written by Matthew J O'Ryan in 1992, provides a snapshot of the collapse and reorganization of horticulture production in New Zealand following the state's reorientation towards market interests. Small orchards across the study area (see Figure 1.11) were becoming increasingly "pluriactive" (O'Ryan, 1992, p. 29) as they sought multiple income streams to allow horticulture to continue. This could be a symptom either of imminent subsumption of peri-urban horticulture to more profitable financial land uses that were already making their presence known in Kumeu-Huapai by the 1980s, or it could be a demonstration of new strategies being adopted to ensure these operations survived in the face of urbanization. O'Ryan himself contended that these orcharding families seeking other forms of employment to bring money back to invest in inputs for horticulture was a clear demonstration of their commitment to the orchard, regardless of what a more calculative and abstract perspective would have to say about such a strategy.

This sentiment from growers 30 years ago has also been observed in the late 2010s in Pukekohe, Auckland's southern urban frontier and one of the most

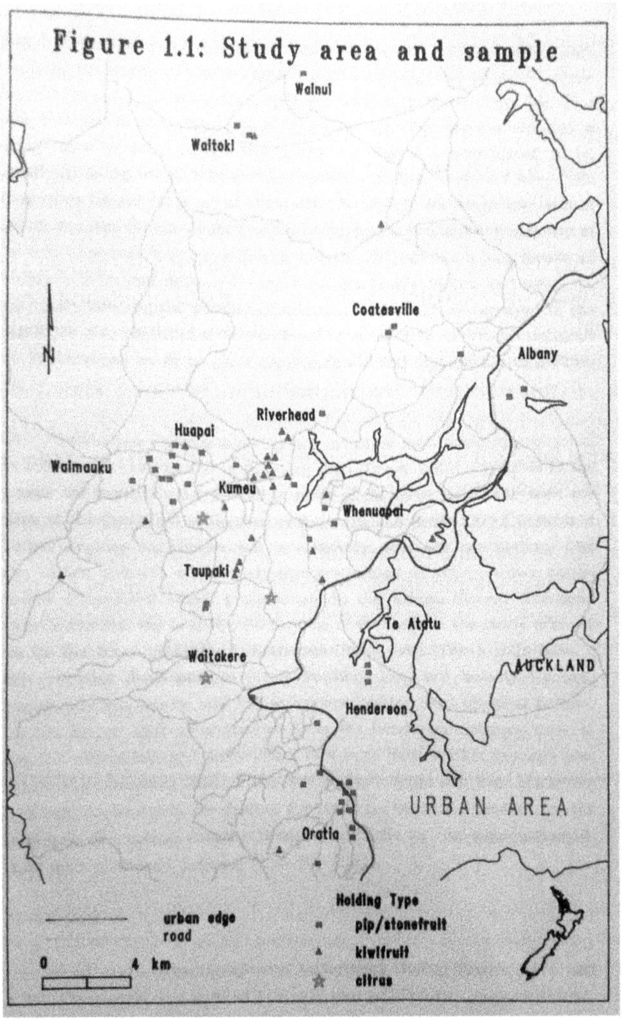


*Figure 1.11* O'Ryan's study area for his 1992 master's thesis fieldwork (p. 8). Note the cluster of pip/stone fruit orchards west of Kumeu-Huapai and kiwifruit orchards east of Kumeu-Huapai.

productive horticultural areas in the country today (Curran-Cournane et al., 2016). Just as their northwestern equivalents did so decades in the past, horticulturalists in Pukekohe are currently facing twin pressures: the mounting difficulties of growing and selling produce at Auckland's periphery and the skyrocketing financial value of their land, should it be converted into residential development. When asked about whether they intended to keep working on their orchard, their concern was not maximizing profitability but simply being able to continue operating (p. 244). For many small growers, the productive work of the household and the relationships among the local horticultural community were more attractive than financial gain. Survival strategies often relied on engaging with abstract financial logic, however, and could in turn lead to the results financially motivated landowners desired in the first place. O'Ryan found that half of the orchardists operating in Auckland's northwest in the early 1990s sought to subdivide part of their land and use the cash to keep producing fruit on a smaller plot of land, many only being held back from doing this by zoning regulations (O'Ryan, 1992, p. 77), leading him to point out that land fragmentation as a survival method would only further entrench urbanization in productive areas (p. 79). In this way, in Kumeu-Huapai 30 years ago and in Pukekohe today, the demands of abstract financial profitability over land and labour eventually erase less "competitive" land uses such as small-scale local food production to serve an urban population.

Exceptions to this rule exist. Two horticultural operations, both run by Croatian families, continue a kilometre or so east of Kumeu-Huapai, on a "golden triangle" of good growing land, but they do so in a markedly different environment to one the older members of each operation remember. "There's nothing left" one grower tells me as we sit in a small packing warehouse adjoining the roadside retail storefront, "everyone jumps to the supermarkets tune". With little in the way of an organized collective front, horticulturalists find themselves outmatched at the negotiating table by large buyers such as New Zealand's supermarket duopoly. As seen in food supply chains across the world (Shaw & Shaw, 2019, p. 26), the supermarkets are free to demand certain quantities, to present prices without negotiation, and to accept or reject produce based on their own standards. At the kitchen table of the home adjoining one of their orchards, the patriarch (or "the old uncle", according to his own estimation) of the family that runs the other extant operation reflected:

> growers had more control [*under the A&PB*] but then they threw in the towel … the board collapsed and the supermarkets swooped in, and all the growers ended up in a worse position … most of the world [*horticulture*] is controlled by big markets, you gotta make a deal with them … they don't want you to die, of course. They keep your head above water, just.

He believes the only path forward was a more calculative approach, and so he has put the business in the hands of nephews and cousins with accounting and finance experience "one worked for Deloittes over in London … it needed revitalizing, we needed a younger and more aggressive bunch".

Both operations are also seeking to make an exit, with very different motivations. One has realized that their children have careers of their own, and without anyone to hand the orchard on to, consider their own retirement to be the end of the enterprise. "I'd like to hand down the orchard, but it won't happen" they say, and then turn their head to study a corrugated wall panel behind me, saying nothing more on the subject. The other, seeing massively profitable developments on rural land occur within view of their own orchards, sought to work with a developer to have their land turned into an aged care facility, directly opposite a recently built childcare centre. Unlike those more connected landowners who are able to work with state and market interests to rewrite the rules and make the greatest profit, this orchard owner was flatly refused their bid to build a rest home on productive land and was left to fume over their lost chance to leave behind the land they had toiled on for decades with an eight-figure sum in hand. "That's not right, if the government, they want to protect soils, that's fine by me, but they aren't being consistent, look what they're doing up there!". As the structures that made peri-urban horticulture possible in Auckland disappear, as subsidies and collective agreements are removed, as orchardists find that work is ever more difficult without unpaid reciprocal help from a community of other growers, those few who remain in the third decade of the 21st century must look to property development for the future of their land, or simply not envisage a future at all.

When current and former horticulturalists looked back on comparatively better conditions for their trade, some continued to hold that the liberalization and financialization of the industry and land had been, if not positive, then at least natural. Either pragmatically ("It was a false economy, it was bound to happen") or regretfully ("it was a false economy, but it was how we made a living"), I was told that the producer-state-market arrangement that allowed fruit growers in Kumeu-Huapai to prosper was a distortion of reality. Kumeu-Huapai's relationship with international finance capital is indeed a long-standing one, and this will be discussed in detail in the following chapter, but one of the local historians I spoke to (again also a former horticulturalist like so many older or retired professionals in the area) took issue with the characterization of the collapse of peri-urban horticulture as a correction of a mistake in some grand system:

> What Brierly's did, was buy an orchard, and when orchardists were scratching around for a living, they said we'll buy shares at such and such price … corporates taking profits they made [*in land and privatization*] and bringing it to places where you can't compete with them … in Australia, they're buying big chunks of water and they're holding them … those are true false economies.

This discursive conflict among people who both lost and gained from such a radical change in the economic order is a particularly clear example of Bourdieu's concerns about the naturalization of the dominant economic order and the "amnesia of genesis" that can come from an analysis that excludes social relationships from economic analysis. To say that the efflorescence of property development in exurban Auckland at the expense of small-scale horticulture is not only an inevitable historical process but also a truer expression of the natural order obscures the very intentional work done to bring this reality about. As described earlier, oral narratives from those who were there to see it characterize the destruction of centralized produce circulation specifically as a bid by a few interested parties – larger industrial-scale producers and large buyers such as supermarkets and financial speculators – to change the political economic order and benefit in doing so. It was, as Harvey puts it, "Ruling Interests, not Ruling Ideas" (Harvey, 2021) that pushed the state to change its relationship to peri-urban horticultural land use in favour of more financial and abstracted values for land. The following and final section of this chapter addresses this significant political project driving liberalization.

## The "restoration" of finance capital

In their respective efforts to dissect and examine the historical alignment of state and capital interests commonly referred to in political or academic parlance as neoliberalism, David Harvey and Pierre Bourdieu each sought to depict the phenomenon as a "restoration of class power" (Harvey, 2007, p. 29), a programme for the enrichment and valorization of capitalists (Bourdieu, 1998), rather than a natural evolution of social life to which no alternative could exist. As with life under any other political project, there is an ideological component to this that is not directly attached to the individual materially gaining from such a state of affairs. The ideas of the dominant class have become, as they say, the dominant ideas (Marx & Engels, 1976), as demonstrated by the former horticulturalists who were bitter at the loss of their previous livelihood yet looked back on that time as an economic anomaly that had been corrected by the forces of history. Yet, patterns of suburban development in peri-urban Auckland are explicitly defined by the demands of finance, the "absolute reign of flexibility" (Bourdieu, 1998, p. 3) in zoning land, designing forms of housing, contracting labour, purchasing materials, all seeking to remove barriers to capital accumulation through the development of rural Auckland. The alignment of interests that allowed for the existence of small-scale peri-urban horticulture required precisely such barriers: a regulated pricing regime, a single authorized circulator of produce for commercial production, state-managed subsidized transport, and (despite the wishes of many growers seeking to sell off a chunk of their land to bring in much-needed cash flow) inflexible regulations on land use and refusals to allow fragmentation through subdivision. Today, the speculative financial value of land is included in investor offerings for some horticulture operations in New Zealand. The idea is that at some point in the future, the profit potential of new suburban housing will overshadow the productive work and, as I heard it put once "grow houses instead of kiwifruit" giving a significant boost to profit figures that would normally reflect the seasonal and not massively profitable work of producing and circulating fruit and vegetables (Fairbairn, 2020, p. 59). The end of horticulture in Kumeu-Huapai, only 20 minutes' drive from Auckland City, can only be clearly depicted by understanding this change in governance and in attitudes towards the value of land.

To look at the business of peri-urban or exurban property development in Kumeu-Huapai and across Auckland today, with its command of massive quantities of materials and human labour, its unmatched speculative financial potential, its seemingly infinite demand for land, is to see an already powerful collection of industries and interests reach new heights of influence and relevance in New Zealand's social order. Representatives of large land developers and housebuilding companies work closely with government-controlled entities to deliver housing at market prices (Gibson, 2021), farming land at Auckland's urban periphery has the potential to sell for a million dollars per hectare (Radio New Zealand, 2019), and the city's unprecedented housing bubble is being watched as a "canary in the coal mine" for the entire global financial market (Cann, 2022).

This set of circumstances is not unique to Auckland or New Zealand – a later chapter will explicitly discuss two international cases to provide comparative context – and it has not occurred through some natural non-directed process of the evolution in governance. Just as a real estate agent in Kumeu-Huapai with a position in local government saw their opportunity to make a good deal for their customers and themselves in the reorganization of Auckland Council in 2010, the frenetic pace of urban development in formerly rural towns exists because suburban development has been encouraged just as less profitable ventures like market gardens close to the city have been deprioritized. "You can't help but bring politics into this, because these are political decisions" one community advocate remarked in an interview, noting "impotency and a feeling of not knowing how to approach this" from residents dealing with the adverse effects of suburbanization, a feeling heightened by the fact the speculative absentee landowners who have bought up rural land around the town have little interest in attending community meetings and calling for reforms: "land bankers aren't chomping at the bit for change".

These local landowners have been working with developers and other interested parties to make money from the accelerating expansion of Auckland for decades, taking their money to retire in comfort. "That retirement village has a lot of local people who sold up" one local government official told me, pointing to the brand-new collection of villas and apartments built on the Huapai Triangle and managed by Hopper Living, who, of course, are family friends of the original directors of Cabra. The changes in policy that rapidly redefined working life in Kumeu-Huapai and many other sites of peri-urban horticulture were, much like many other changes that took place in political and economic life in the last quarter of the 20th century, very deliberate attempts to recentre financial profitability for holders of capital as the organizing principle for life (Gudeman, 2016, p. 164). People who stood to gain from this revolution have supported these policies as they were being introduced, or like the horticulturalists who find themselves without the means to carry on their work, eventually accepted the payout offered for the speculative value of the land they had worked, "grumbling all the way to the bank", as one interlocutor put it. The changing shape of work and life in Kumeu-Huapai and other formerly rural towns at Auckland's urban periphery is an expression of a move away from the productive base of the economy in favour of more profitable financial values to the benefit of financial interests and landowners.

This chapter has contended that the most dramatic aspects of Kumeu-Huapai's suburbanization – rapid pace and massive profitability through residential development on former growing land, a reorientation of social life in the area towards Auckland City, the disappearance of local horticulture – as symptoms of a historical reorientation of land use governance to more explicitly financial goals over less profitable or competitive land uses and that this reorientation benefitted those who supported and enabled it. Just as this state of affairs is historically constructed, so too was that vanished formation of production and state interests that allowed peri-urban horticulture to thrive at the edge of a major city. To examine the factors that created and eventually destroyed that social order, the next chapter takes a long view at the history of the Kumeu River Valley and the people who live there, from before the time of its colonization by the British Empire to today, examining how successive waves of capitalization of the land in different forms sought to exploit or reshape what had come before. As the "restoration" in Harvey's title for the neoliberal era implies, what is happening around growing cities is not without precedent. Far from being a new arrival to Kumeu-Huapai, forms of state governance subservient to financial interests have sought to mould the lands around the Kumeu River since before many Europeans had ever set foot there.

## Bibliography

ACC. (2022). *Injury Prevention, Workplace Safety: Tackling the Hidden Cost of the Building Boom.* https://www.acc.co.nz/newsroom/stories/tackling-the-hidden-cost-of-the-building-boom/

Arrighi, G. (2010). *The long twentieth century: Money, power, and the origins of our times.* New York: Verso.

Auckland Council. (2016a). *Auckland Council GeoMaps.* https://geomapspublic.aucklandcouncil.govt.nz/viewer/index.html

Auckland Council. (2016b). *Special Housing Area Map (Tranches 1–10).* Retrieved January 31 from www.aucklandcouncil.govt.nz/grants-community-support-housing/Documents/sha-overview-tranches-one-to-ten.pdf

Auckland Council. (2019). *Rodney Local Economic Overview 2019.* Retrieved April 27 from https://www.knowledgeauckland.org.nz/media/1841/rodney-local-economic-overview-2019.pdf

BDO. (2022). *Rethinking Construction: 2021 BDO New Zealand Construction Sector Report.* https://www.bdo.nz/en-nz/construction-survey

Bourdieu, P. (1998). The Essence of Neoliberalism. *Le Monde diplomatique.* Retrieved December from https://mondediplo.com/1998/12/08bourdieu

Bourdieu, P. (2005). *The social structures of the economy.* Cambridge: Polity.

Cann, G. (2022). *World Watching NZ Housing Market as Auckland Labelled 'Canary in Coal Mine'.* Retrieved April 28 from https://www.stuff.co.nz/business/128396590/world-watching-nz-housing-market-as-auckland-labelled-canary-in-coal-mine

Centrix. (2021). *National Credit Limit Exceeds $300 Billion.* Retrieved April 27 from https://www.centrix.co.nz/national-credit-limit-exceeds-300-billion-up-28-billion-annually/

Curran-Cournane, F., Cain, T., Greenhalgh, S., & Samarsinghe, O. (2016). Attitudes of a farming community towards urban growth and rural fragmentation—An Auckland case study. *Land Use Policy 58*, 241–250.

Curran-Cournane, F., Carrick, S., Barnes, M. G., Ausseil, A.-G., Drewry, J. J., Bain, I. A., Golubiewski, N. E., Jones, H. S., Barringer, J., & Morell, L. (2021). Cumulative effects of fragmentation and development on highly productive land in New Zealand. *New Zealand Journal of Agricultural Research*, 1–24.

d'Amour, C. B., Reitsma, F., Baiocchi, G., Barthel, S., Güneralp, B., Erb, K.-H., Haberl, H., Creutzig, F., & Seto, K. C. (2017). Future urban land expansion and implications for global croplands. *PNAS, 114*(34), 8939–8944.

Dunsford, D. (2002). *Doing It Themselves: The Story of Kumeu, Huapai and Taupaki.* Auckland: Kumeu District History Project.

EBOSS. (2021). *2021 Construction Supply Chain Report.* Retrieved from https://www.eboss.co.nz/construction-supply-chain-report/q3-2021/background

Fairbairn, M. (2020). *Fields of gold: Financing the global land rush.* Ithaca: Cornell University Press.

Fernandez, R., & Aalbers, M. B. (2016). Financialization and housing: Between globalization and varieties of capitalism. *Competition & Change*, *20*(2), 71–88.

Gibson, A. (2021). *NZ's biggest builder wins consent to buy 20ha Riverhead site for vast housing project*. Retrieved April 28 from https://www.nzherald.co.nz/business/nzs-biggest-builder-wins-consent-to-buy-20ha-riverhead-site-for-vast-housing-project/CUTPHKQ3QPMABAQYJA3S6CTYEI/

Google LLC. (2022). *Google Satellite via QGIS*. Retrieved from http://mt0.google.com/vt/lyrs=s&hl=en&x={x}&y={y}&z={z}. Source: http://mt0.google.com/vt/lyrs%3Ds%26hl%23Den%26x%23D%27Bx%27D%26y%23D%27By%27D%26z%23D%27Bz%27D&zmax=18&zmin=20. Provider: wms.

Gudeman, S. (2016). *Anthropology and economy*. Cambridge: Cambridge University Press.

Harvey, D. (2007). Neoliberalism as creative destruction. *The Annals of the American Academy of Political and Social Science*, *610*, 22–44.

Harvey, D. (2010). *The enigma of capital: And the crises of capitalism*. London: Profile.

Harvey, D. (2021). [S3 E17] *The Ruling Ideas of a Billionaire Class in the Anti-Capitalist Chronicles*. Retrieved from https://youtu.be/0pHfgFVt-XE

Kelsey, J. (2015). *The FIRE Economy: New Zealand's Reckoning*. Wellington: Bridget Williams Books.

Kumeu Central Property Limited. (2022). *Kumeu Central*. Retrieved May 8 from https://kumeucentral.co.nz/

LINZ Data Service. (2020). Land Information New Zealand. Retrieved May 28 from https://data.linz.govt.nz/

Marx, K., & Engels, F. (1976). *The German ideology*. Moscow: Progress Publishers.

Ministry for the Environment and Stats NZ. (2021). *Our Land 2021* (New Zealand's Environmental Reporting Series, Issue). Retrieved from https://environment.govt.nz/publications/our-land-2021/

Ministry of Business Innovation and Employment. (2022). *Construction Sector Trends Biannual Snapshot*. Retrieved November 2021 from https://www.mbie.govt.nz/building-and-energy/building/building-system-insights-programme/sector-trends-reporting/biannual-snapshots/construction-sector-trends-quarterly-snapshot-november-2021/

Murphy, L. (2019). Financiers and developers: Interviews concerning their interests, relationships, and the residential development process. *Building Better Homes Towns and Cities SRA 1 – The Architecture of Decision-making Working Paper*, March 2019.

Murphy, L. (2020). Performing calculative practices: Residual valuation, the residential development process and affordable housing. *Housing Studies*, *35*(9), 1501–1517.

Oneroof. (2020). *225,000 NZ Homes Earned More than Average Kiwi*. Retrieved from https://www.oneroof.co.nz/news/225000-nz-homes-earned-more-than-the-average-kiwi-37599

O'Ryan, M. J. (1992). *Orcharding in the urban periphery: An examination of family orchard management strategies in Auckland's Northwestern Periphery*. Auckland: The University of Auckland.

Radio New Zealand. (2019). *Building on Prime Growing Land – Nine to Noon*. Retrieved May 2 from https://www.rnz.co.nz/audio/player?audio_id=2018693232

Reserve Bank of New Zealand. (2021). *House Prices and Values*. Retrieved from https://www.rbnz.govt.nz/statistics/key-graphs/key-graph-house-price-values

Shaw, H. J., & Shaw, J. J. (2019). *Corporate social responsibility, social justice and the global food supply chain: Towards an ethical food policy for sustainable supermarkets*. London: Routledge.

Stats NZ. (2020). *Place Summaries: Kumeu-Huapai*. Stats NZ. Retrieved June 4 from https://www.stats.govt.nz/tools/2018-census-place-summaries/kumeu-huapai

Stats NZ. (2021). *Building Consents Issued: June 2021*. Retrieved September 6 from https://www.stats.govt.nz/information-releases/building-consents-issued-june-2021

The University of Auckland. (2020). *University of Auckland Human Participants Ethics Committee (UAHPEC) Applicants' Reference Manual*, June 2020. Retrieved April 27 from https://www.auckland.ac.nz/en/research/about-our-research/human-ethics/human-participants-ethics-committee-uahpec/essential-reading.html

Wetzstein, S. (2021). Toward affordable cities? Critically exploring the market-based housing supply policy proposition. *Housing Policy Debate 32*, 1–27.

# 2 The historical process of the capitalization of the Kumeu River Valley

*Figure 2.1* A massive flood overtakes the colonial settlements of Kumeu and Huapai nearly a century ago. This picture was published by Deborah Dunsford for her history of the area, and it is reprinted here with her permission (2002, p. 116).

"It's part of growing up", Derry tells me. "Those Chinamen growing cabbages in Rosebank Road, it didn't matter how good the cabbages were, the dollar wins". Derry Faulder is a local historian and long-time resident of the area. He has been living and working here long enough, in fact, that I found in our first interview that he was apprenticed as an electrician to my maternal grandfather, who had passed away when my mother was young. As we sat and talked, either at the meeting room of the Kumeu Agricultural and Horticultural Society or the dining table of his home just north of Kumeu-Huapai, Derry would freely jump from the topic of urban development and changing patterns of life in the area to anecdotes about my grandfather that he would use to reinforce his point. "Make the best of the era you're in, give education to the children, and move on. Bill did the same". As he spoke

DOI: 10.4324/9781003400325-4

of the origins and assimilation of the Croatian community to which Bill Orsolich belonged "Dalmatians and Dutch came here as migrants, peasants fleeing from war and famine" at the start of the 20th century, and the accelerating advance of urban social and economic life onto the periphery "expansion gobbles up a Nelson or a Rotorua every two years",[1] he would clutch his hand to his head and clench his jaw for a while before speaking, as though the act of summoning the images of a time half-a-century past and placing them alongside current events and relationships took a great deal of physical effort. For Derry, the process that created his current house and its one-acre block of land out of a 250-acre farm is as natural as the tides and the process of aging, and attempts to forestall or deny it have an equal chance of being successful. "The residents of Mt Albert probably said all the same things in 1896", referring to the fate of places like Rosebank Road and Mount Albert – former horticultural areas closer to Auckland City – as case studies for the inevitable urbanization of formerly rural towns like Kumeu-Huapai today.

This apparently endless process was depicted as a series of waves emanating out from the centre of the city over the course of more than one-and-a-half centuries of colonial development, as people moved out repeatedly from the city centre over generations. Livestock farmers and horticulturalists (such as Derry's own family) sold land to housebuilders and put the proceeds towards buying land further away from the city. A common pattern I heard in other interviews regarding this slow migration was that it began in what is now Auckland's inner suburbs, then moved to Rosebank Road and Te Atatu, then Lincoln Road in Henderson, then Whenuapai and Kumeu-Huapai, and, perhaps in the future, more distant areas north of the City (see Figure 2.2). "There'll be a mall right here, someday" Derry says, gesturing around at his house, then at the lifestyle blocks and the commercial pine forest visible in the windows north of us "it takes a long time, but it will happen".

For many who live in Kumeu-Huapai, the rapid pace of suburban forms of residential development over the last two decades is understood to be another stage of the life cycle of rural land at the periphery of a growing city. The previous chapter demonstrated that this naturalization of development can hide the very intentional actions of interested parties seeking to make the highest possible profit from the financialization of land, but another side of this discourse is whether or not Kumeu-Huapai's transformation of land uses towards explicitly financial ends is a new occurrence in its history. Many older residents and former horticulturalists recall a time when life and work were more situated around local socialization and production, yet this chapter examines the successive forms of colonization and capitalization that have shaped the Kumeu River Valley in order to argue that the period of time where peri-urban horticulture was a dominant way of life in the area was not some bucolic colonial origin onto which urban life has finally intruded, but an interregnum in the financial order, ended by the "restoration" of financial power in the latter quarter of the 20th century. Indeed, when some former horticulturalists in the previous chapter saw the loss of their old way of life as a correction in the distortion of global markets, such a view of financial power over colonial peri-urban land may indicate a very deep understanding of the history of the valley, as much as it can be seen as evidence of the power of the dominant ideology.

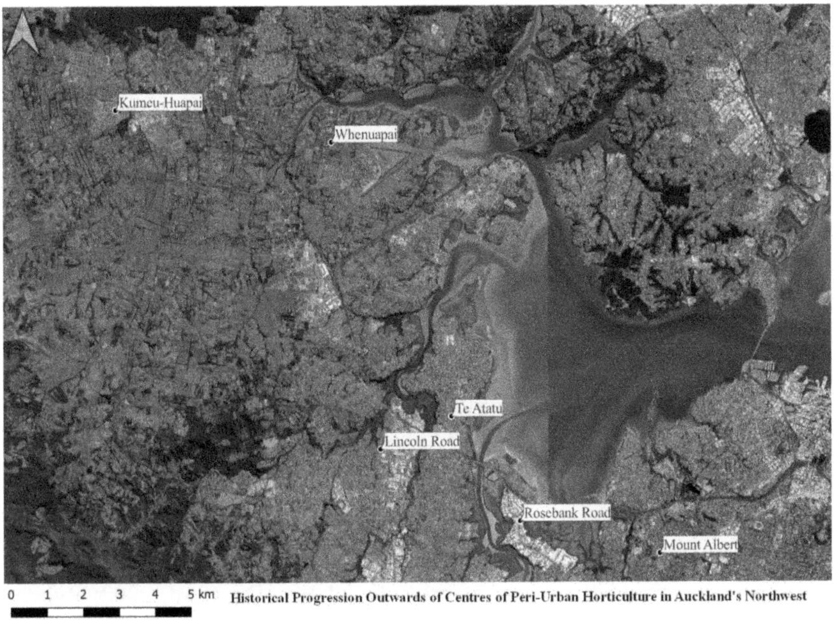


*Figure 2.2* Centres of horticultural production in Auckland's northwest mentioned in interviews, moving east to west over time. Markers added by author, satellite data from Google Maps (Google LLC, 2022).

## ***Expectations of modernity* and an anti-teleological approach to historical narratives**

In seeking a framework to examine the history of such a financially lucrative area, a formerly rural town that recently commanded the most dramatic rises in property values seen anywhere in a country internationally notable for high house prices, I turned to James Ferguson's (1999) study of social and economic life at the bottom of the global imperial order. Ferguson charted the "brutal and blinding speed" (p. 2) of Zambia's industrialization and "modernization" over the middle of the 20th century, a rapid ascent to the forms of production that western academics and many people living in Zambia believed heralded European-style wealth and power for the country (p. 5) before a sudden shift in national fortunes seemed to suggest a reversal of linear history, a falling away from a promised urbanized future towards a move "back to the land" (p. 11). His analytical approach to this problem of failed narratives and unmet expectations of modernity is to specifically avoid the history of the Zambian Copperbelt as a linear teleological progression of types, instead presenting t he processes that brought about the country's rapid rise and fall in the global order as a shift in the relative arrangement of a collection of tendencies, arguing "changing realities must be conceptualized not as ladders or trees defined by sequences and phases but as dense "bushes" of multitudinous coexisting variations, continually modified in complex and nonlinear ways" (p. 42). To this end, I will present the history of the capitalization of the Kumeu River Valley as substantively as can be done from a historical perspective, referring to the expropriation and conversion of specific blocks of land and the actions and intentions of individuals, rather than the river valley's place as part of a broad formalist story of Aotearoa's conquest at the hands of the British crown and its colonization by European settlers over the following centuries. By taking such a close-up view of the last 200 years of life in this one place at the periphery of Auckland, I intend to demonstrate the relationships between people, land, and capital accumulation that made Kumeu-Huapai today.

A complication to this ambition is the paucity of sources on the Kumeu River Valley as anything but an Anglo and Dalmatian colonial settlement. One member of Ngāti Whātua O Kaipara's development trust[2] explained it so:

> The impact of the expropriations of course is that our people have become invisible in the area and so have their old associations, and thus disconnected. Many people never understood how that happened, so the Waitangi Tribunal process was a bit of a revelation.

Because of this, my historical perspective relies on a few local histories that preserve some remaining fragments of pre-colonial life (Dunsford, 2002; Ryburn, 1999) and testimonies of expropriations of Māori land in the Kumeu-Huapai area bought before the Waitangi Tribunal for the purposes of beginning the process of restitution (Stirling, 1998). This latter collection of oral histories and mid-19th-century archives is a very good source for learning how exactly land came to be possessed by public or private colonial entities, and the official histories of the Kaipara region and of Auckland's northwest do contain some insights into the capitalization of Kumeu-Huapai's land, but I must work with a far less complete set of information than the previous chapter, where I could rely on the testimony of dozens of professionals and locals with direct experience of the process today.

### "*Kumeu*: lit. to drag/pull something + breast, as in a place to feed"

The placement of the Kumeu River Valley between the Waitemata Harbour to its southeast and the Kaipara Harbour to the northwest made it an important portage (see Figure 2.2) for Tangata Whenua centuries before Europeans were attempting to make the journey between Auckland city and the west coast. Kumeu itself could not be described as a permanent settlement, at least according to accounts that survive today. This fact was leveraged by some interlocutors to claim the area had never been occupied by Tangata Whenua and was thus explicitly a European settlement, the phrase "there's no Māori's here" coming up more than once. As its name implies, Kumeu was a rest stop on the road between Tamaki Makaurau and the Kaipara. As a crucial avenue of trade and transport, the river valley saw migration and conflict, particularly the gradual displacement of the Kawerau peoples by Ngāti Whātua from approximately 1250 ce onwards, to such a degree that they were almost completely wiped out by the time of European arrival (Dunsford, 2002, p. 15). As the Musket Wars in the first quarter of the 19th century, conflicts of

Tikanga (custom) spurred on by the use of European forms of food production and military technology (Ballara, 2003) drastically reshaped relationships between iwi, and Ngāti Whātua found themselves pushed south and into the Waitakere ranges by Ngāpuhi just as they had displaced the Kawerau. In 1832, a group of Anglican missionaries walking from Kaipara to Waitemata and then on to Manukau (today in the south of Auckland) claimed to not see a single cooking fire for the entirety of their journey (Dunsford, 2002, p. 17).

European settlers moved to fill the space left by conflict rapidly, and a decade later, there were approximately 3,000 colonists living in the Auckland area. Seeing a competitive advantage, Ngāti Whātua began to grow and sell produce to the Europeans. By the middle of the century, Reweti (southeast of Helensville in Figure 2.3) provided potatoes, cabbages, and onions to Auckland (p. 21). By this same time, the colonial government's main method of financing development – buying land from Māori and selling it to settlers at a profit – had been undermined to allow deregulation and privatization of the process (p. 19). In 1865, with the passing of the "Native Land Act", intentionally designed to remove the concept of shared ownership and encourage private property relations by limiting the number of possible owners of land, described by Sir Hugh Kawharau as a "veritable engine of destruction for tribal tenure" (p. 25), the Kumeu River Valley became an object of interest for land bankers and other financial interests more than one-and-a-half centuries before I would hear complaints about the sudden rise in absentee landlords in the area. Its historical importance as a rest stop on the trail to and from Tamaki Makaurau/Auckland led capitalists – first those working directly under the colonial project, then private individuals – to seek ways to put their money into Kumeu for a profit even as the initial annexation of Aotearoa by the British Empire continued to be fought out.

## Deacon, Davis, and three separate men called William Deloitte: the expropriation and financialization of the Kumeu River Valley

The very moment a narrative of Kumeu-Huapai moves from the days before private expropriation of iwi land to the times it began to be seen in the area, sources immediately become more detailed, and individual agents and blocks of land suddenly become much more easily discerned. The Kumeu portage itself was purchased by the colonial government, but the area that today comprises much of Kumeu-Huapai and the neighbouring formerly rural towns of Waimauku and Taupaki had not been bought by the Crown before the process became privatized. Because of this, the land northwest of Auckland began its colonial life as a series of blocks (see Figure 2.4) purchased from Ngāti Whātua under extremely suspect circumstances. Utilizing finance as a weapon to subjugate one's enemies is a particular form of conflict that David Graeber associated with the imperial expansion of Occidental/Christian/European power to such a degree that he asserted that a direct association between commerce and domination was a unique feature of Occidental/Christian Ideology (Graeber, 2011, p. 291), and the lands of the Kumeu River Valley saw much of this form of colonial warfare. Thomas Deacon and his

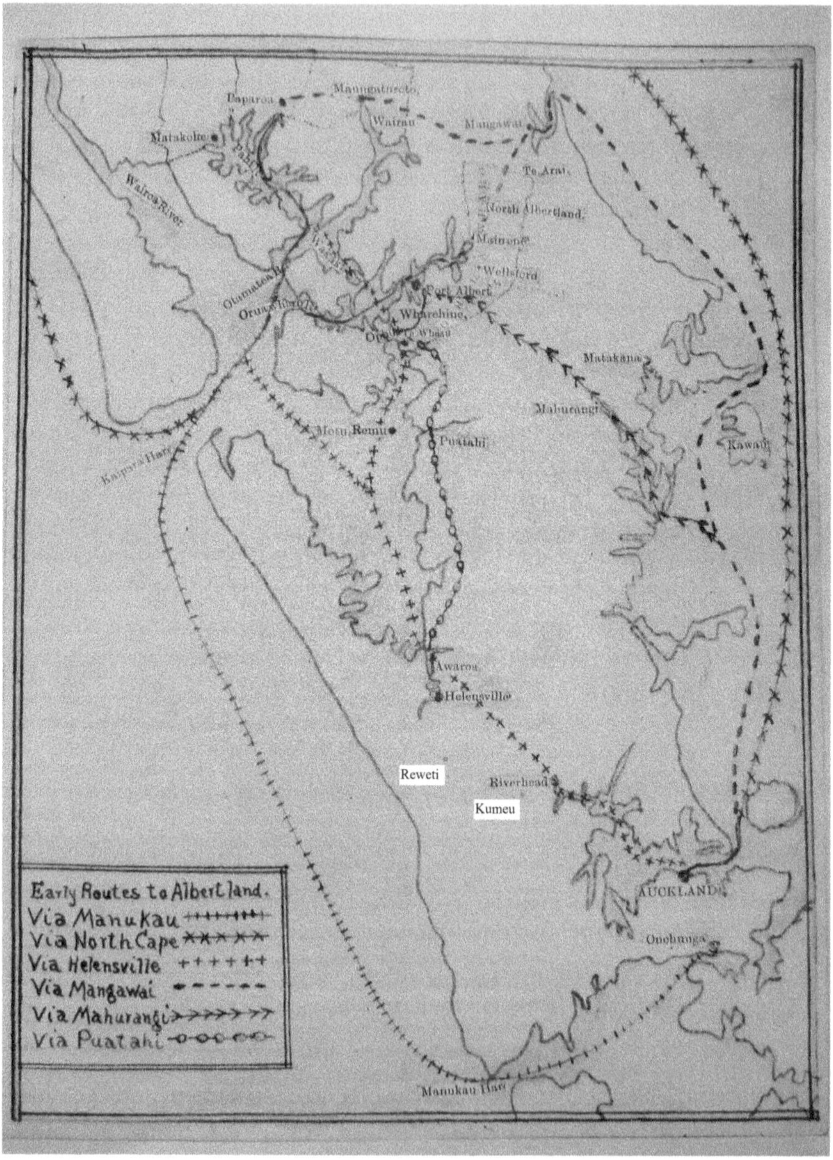


*Figure 2.3* Map of routes in the 1860s from Auckland to "Albertland" in the North Kaipara Harbour (Ryburn, 1999, p. 17). Kumeu and Reweti have been marked on the map by the author. Reprinted here with the permission of the Albertland Heritage Museum.

family ran a storefront and accommodation around the Kumeu portage, and very quickly, they began to offer goods to Ngāti Whātua on credit with steep interest, taking their land as recompense with the backing of the colonial state (Dunsford, 2002, p. 33). Through this method of enclosure, Deacon acquired the Turakiawatea and Te Ihumatao blocks one by one, personally owning nearly 800 acres of the Kumeu River Valley (Stirling, 1998, p. 436). Meanwhile, Ngāti Whātua hoping to sell the massive Taupaki block found themselves rapidly indebted by unusually high surveying and legal fees levied by colonial agents and faced resistance from European buyers who claimed, as one, that they could not accept any purchasing deal due to the 10 per cent government purchasing fee (Stirling, 1998, p. 282). The selling price was pushed ever lower and debts mounted ever higher until a speculator named William Deloitte bought himself 2000 acres of high-quality land at less than a shilling an acre (p. 282) and drew up mortgages at 10 per cent interest for other, less wealthy agents wishing to buy the Waikoukou block (p. 421).

William Henry Deloitte was the son of William Salmon Deloitte, who first bought European access to Hokianga port (north of the Kaipara) from iwi in 1826 before rapidly selling the land rights on to Sydney merchants (Carter, 2016) and spending the rest of his life as a speculator and apparently passing on his trade. He is, as some readers may have guessed, also the cousin of William Welch Deloitte, founder of the "big four" consulting company, whose skills and services a horticulturalist in the previous chapter claimed would help his operation survive and adapt in the face of urbanization. All three William Deloitte's – the speculator and his father operating out of Sydney and the accountant working from London – are directly descended (William Salmond Deloitte is his son, the other two are grandsons) of Count Jean De Loitte, the treasurer to king Louis XVI who had fled

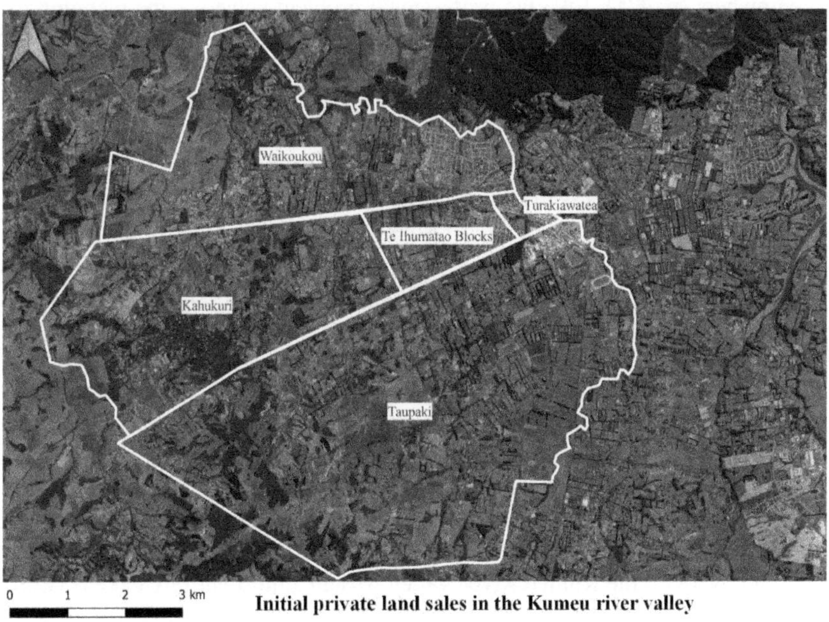


*Figure 2.4* The private sales of land that came to be Kumeu-Huapai and its surrounds (Stirling, 1998, p. 276). All land east of the Kumeu River (the eastern border of the Taupaki, Waikoukou, and Turakiawatea blocks) was a Crown sale. Satellite imagery from Google Maps (Google LLC, 2022).

to England at the onset of the French Revolution (Parker, 1984, p. 140). That the direct descendants of a powerful financial figure would each continue to play an important role in profit-making in the imperial order is something that has clearly not gone away in Kumeu-Huapai, if the dense network of connections between family-owned land development companies, their small selection of specialist professional services, and the government today is anything to go by.

As well as ambitious merchants seeking to enclose the frontier and international financiers descended from the Ancien regime speculating and mortgaging vast swathes of land, the picture of the expropriation and capitalization of the Kumeu River Valley would not be complete without the figure of Charles Oliver Davis, land agent and greaser of the great wheels of colonial project. Simply described as "infamous" in Waitangi Tribunal narratives about his role (Stirling, 1998, p. 279) and hailed in his obituary and official accounts of colonial history as a well-known scholar and interpreter (New Zealand Herald, 1887), Davis had been in Aotearoa for 40 years already by the time of the private land sales in the Kumeu River Valley and had struck deals with hapu and iwi for over a million acres of land across the North Island by the time he made his appearance there (Ward, 1990). Davis removed the children of the Ngāti Whātua land sellers as interested parties to streamline the selling process in the Taupaki block to Deloitte (Stirling, 1998, p. 279), acted as a legal representative to assert the right of the Turakiawatea block to be sold to Thomas Deacon (p. 434), and was the interpreter for the land sale of the Waikoukou block and its subsequent mortgaging by Deloitte to its mortgagee owners (p. 421). Like the professional problem solvers – the planning experts and environmental lawyers, ever on call to a land developer should a project face opposition – that provide specialized and indispensable skills for smoothing the way for capital accumulation in Kumeu-Huapai one-and-a-half centuries later, Davis ensured the first phase of financialization of the river valley.

A final example of the expropriation and accumulation of land in the early colonial history of the area is the buying process of the Kahukuri block, west of the land Deacon acquired (Figure 2.4). An Auckland storekeeper named Bernard Levy and the lawyer John Russell, founder of the law firm Russell McVeigh, one of the three largest legal firms in New Zealand today, offered Ngāti Whātua owners of the Kahukuri block an unusually high rate of 20 per cent per annum for a mortgage on their land in 1870. They then acquired the land from the defaulting owners in 1871, selling it on to farmers at an easy profit (Stirling, 1998, p. 447). Land acquired by Deloitte, Deacon, and almost every other buyer was utilized in the same manner: nearly every purchase was made for the purpose of profitable speculation and sale of subdivisions of the blocks to prospective farmers arriving to make their new life at the colonial frontier (Dunsford, 2002, p. 26). Debts continued to force Ngāti Whātua to sell land to these large buyers who would in turn offer the land to colonists, and by 1877, the Kumeu River Valley was an almost entirely Pakeha (European New Zealander) settlement (p. 38). Thus, the first stage of the capitalization of land around Kumeu-Huapai was one of accumulation by dispossession (Harvey, 2017), alienating tribal land tenure in order to privatize it, subdivide it, and sell it to those hoping to make a profit from working the land in a way or another.

Thereafter, some land in the area was used for livestock or horticultural agriculture, but the place where Kumeu-Huapai stands today was densely forested (and rainfall in the area remains at the level one sees in rainforests to this day, a fact that will become very important in the later chapter on climate change), so the first profitable productive work extracted from the land of Kumeu-Huapai was timber felling, followed immediately by digging for deposits of Kauri gum[3] left in the earth long after the great Kauri trees had been cut down and sawn into planks (Dunsford, 2002, p. 45). With the rainforest gone, the river and the rains were free to turn the river valley into a swamp, particularly in the areas that make up the eastern side of Kumeu-Huapai today, in close proximity to the river and of very low elevation as it is. The new century rose on the river valley's transformation into a patchwork of low bogs and sparsely populated hilly farmland, even as the area continued to be an important rest stop on the route from Auckland to the west and north as it had always been. The first two phases of capitalization over and done with, the river valley would need a new way of valuing and selling such denuded and degraded land to willing and paying customers at a profit. Capitalists would have to turn to the art of advertising.

## Huapai Estate: Auckland's suburban fruit colony, brought to you by Northern Fruitlands Ltd

For the first part of this narrative, only half of Kumeu-Huapai has appeared under its modern naming convention. Kumeu is a pre-colonial title for a portage, a place to rest as one drags their boats from one body of water to another. The name "Huapai", however, was first applied to the land directly west of the Kumeu portage as a marketing effort by a land development company. Northern Fruitlands Ltd was founded in 1914, and the company shares were split between several Auckland-based professionals (Dunsford, 2002, p. 59). Northern Fruitlands bought land from the Kumeu portage westwards, almost reaching the western coast at Muriwai (see Figure 2.5), from landowner families such as the Deacons, as well as other names that now grace local street signs such as Boord and Foster. The land was divided into lots, and the whole development was named after a direct transliteration of the Te Reo Māori words "good fruit". The properties for sale, a mix of 10- and 40-acre blocks, were advertised to Auckland residents as a chance to escape the city and embrace the bucolic splendour and profitable industry of orchard life. Those hoping for a life of leisure were shocked and disappointed to find the material reality of the Kumeu River Valley and the Muriwai coast: the salt-laden westerly winds from the Tasman Sea soured the earth of the hillsides, Kauri logs left behind from the timber operation to sink into the sediment of the valleys and had to be laboriously hauled out, and entire fields of bogs and swampland needed to be drained and removed before the land could even begin to be worked (p. 62). Like so many who would come after them, the first European horticulturalists of Kumeu-Huapai were sold a fiction in order to create profitable land sales.

To this day, the former and current horticulturalists of Kumeu-Huapai have a complicated and sometimes even bitter relationship with the land they worked, even as the labour they put in over decades markedly improved the growing

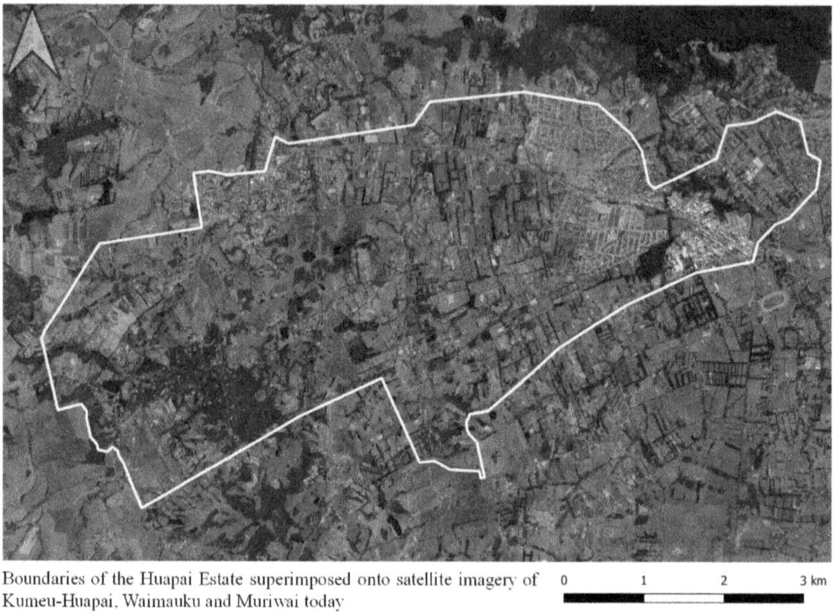


*Figure 2.5* The boundaries of the Huapai Estate sale in the 1910s (Dunsford, 2002, p. 59) overlaid onto the area today. Satellite imagery from Google Maps.

potential of the area. The century and a half of human transformations and geologies that define the land have rendered Kumeu-Huapai a chaotic palimpsest of soil formations (O'Ryan, 1992, p. 9). One soil and land use expert told me "within a property there's often two or three variations" of distinct soil categories, making broad categorizations of Land Use Capability impossible, and a former horticulturalist pointed to one area of land in Kumeu-Huapai (that has since been built on with suburban housing), told me it bore distinct similarities to the extremely rich volcanic soil that helps make Pukekohe in Auckland's south so productive today, and then spoke dismissively of the growing conditions of the land across the same road.

An added difficulty that confronted horticulturalists then and now is the rainforest climate that remains long after the rainforest is gone. Humidity and heavy rainfall make the Kumeu River Valley an inviting environment for pests and parasites, requiring growers to make heavy use of pesticides or other (less toxic) management techniques to protect their produce. Massive flooding events occurred on a generational timescale, the first major inundation sweeping the fledgling township in 1924 (see Figure 2.6) and have struck again every few decades since. Derry remembers wading through a flooded main street in 1954 to seek shelter in the Town Hall, and the flooding of Kumeu-Huapai during my period of fieldwork will be the main focus of the final chapter of this book. Despite these barriers, the sale of the Huapai Estate was a success for the company, and Kumeu-Huapai began to take shape in the way it stands today. By the end of 1921, there was a metalled (gravel) road from Auckland through the settlement, a horticultural co-op had been formed, and the first annual Kumeu Show had taken place (p. 65).

*Figure 2.6* Mavis Barnes, Inez Marshall, and Evelyn Barnes stand in front of the submerged Post Office. Like the image at the start of the chapter, this photo from the 1924 flood was published by Deborah Dunsford (2002, p. 117) as part of her historical research, and it is reprinted here with her permission.

## Dalmatian settlers in Kumeu-Huapai

Around this time, at the turn of the century and onwards, Dalmatian (Croatian coast) immigrants, "the peasants fleeing from war and famine", began to arrive in the Kaipara region. Once, Derry pointed to the rough path of the Kumeu Portage from the Kaipara Harbour to Kumeu visible north of his house, turned to me, and said, "that's where your lot came from". I have no connection at all with the Croatian community of Kumeu-Huapai or the wider Auckland region, the early death of my maternal grandfather and my maternal grandmother's decision to remarry along with my mother's side eventually moving away from the area all contributed to the fact I learned significantly more about my grandfather and the Dalmatian community in Kumeu-Huapai from this research project than I had encountered in the rest of my life put together. Like most European New Zealanders, my immediate ancestors on my father's side and half of my mother's are Scottish, English, and Irish, workers and storekeepers who have formed the core of British imperial expansion across the globe for centuries. Yet, Bill Orsolich, this single long-passed grandparent, possessed an outsized importance in conversations with some interlocutors as a representation of life in Kumeu-Huapai over the first half of the 20th century. "He marry a [*British*] New Zealand girl?" one of the remaining horticulturalists asked (as mentioned in the previous chapter both extant operations are owned by Croatian families), then he shook his head and explained there is a marked difference between those who integrated into Anglo colonial life and the families like his that continued to marry within the diaspora, "some people kept the ties going".

Derry used Bill's work as an example of this social separation between settler communities, explaining that he had to learn some words of Serbo-Croat in his apprenticeship as an electrician in order to work alongside Bill when they worked their trade across the river valley, because the Dalmatian customers would (mostly) only speak their language and offer work to fellow Croatians. The British majority reacted poorly to a possible competing ethnic group in Auckland's northwest and resented the fact that Dalmatian migrants would utilize a wide network of kinship relationships to put together the money to buy land or the labour to run a gum-digging team, rather than relying on the work of a nuclear household for production and social reproduction (Dunsford, 2002, p. 152). Accounts of colonial life in the Kaipara from the turn of the century bear British descriptions of Croatians as swarms of locusts, mobs, or invasive pests, alongside other "black" immigrants from China, India, and the Levant (Bozic-Vrbancic, 2006, p. 186), but by the 1930s Croats were landowners in Kumeu-Huapai, pioneering the wine industry in the area and growing a wide range of fruit and vegetables for consumption and sale (Dunsford, 2002, p. 159). To this day, the family names behind Kumeu-Huapai's wineries – Nobilo, Selak, Brackovich, and Soljan – and the fact that both remaining horticultural operations are run by Croatian families point to the outsized role Dalmatian immigrants have played in the development of peri-urban horticulture in Kumeu-Huapai.

## The "False Economy": mid-century interregnum and peri-urban horticulture

The dire economic consequences of the Great Depression and the vast national effort demanded by the Second World War each inspired the creation of government-controlled production systems such as the Boards and encouraged a general posture of the state in favour of protecting small-scale peri-urban horticulture (Dunsford, 2002, p. 179). It is this time period, now fading from living memory, which holds that essential place in the minds of some current and former horticulturalists as the lost golden days of the trade and to others as an interventionist distortion that could never have survived. Urbanization closer to the city prompted successive waves of movement outwards from horticulturalists in Rosebank, Te Atatu, and Henderson, who then bought 10- or 40-acre blocks first drawn up by the Huapai Estate development (see Figure 2.7) or subdivided from larger farms. While the actual acreage of orchards only declined after the 1920s, as the initial attempts by settlers to turn the deforested bog sold to them by Northern Fruitlands into productive land met with very variable results (p. 179), the middle third of the century saw Kumeu-Huapai's efflorescence as a productive centre to Auckland City. Orchardists planted fruit trees, and in the years it took for the orchards to mature, they used the land for horticulture cropping, "as the orchards developed, everyone grew crops, squash, tomatoes ... gave the whole area a sort of foodbowl" a local historian explained. One former horticulturalist, after despairing at how quickly their work had not only collapsed but become invisible in the new landscape of the urbanizing town, made a point of telling me that Kumeu's apple orchards had been visited by Queen Elizabeth II on her coronation tour across the commonwealth in 1953, “we were the biggest in Auckland back then”. They remember how they’d “come down on a winters night and unload thirty cases of grapefruit” onto the train, leaving an extra case for rail staff moving the produce across the country so they wouldn’t help themselves to the main stock of produce.

Land banking and speculation were never entirely absent from land use in this period. Even constrained by more rigorous subdivision rules, the steady urbanization of New Zealand’s population made ownership of peri-urban land a speculative asset. Many orchard owners in the mid-century heyday were in fact Auckland businessmen, “white-collar orchardists” who had seen an opportunity to diversify their time between urban professional life and horticultural production. Instead, the tendency to financialize the land of the Kumeu River Valley that defined its colonial history before and after the middle of the 20th century was rivalled in this period by the attitude of the state and the co-operatives of small horticulturalists towards the productive value of peri-urban growing land. This particular combination of historical tendencies and alignment of interests encouraged small-scale horticulture at the margins of Auckland City. Whether interlocutors saw this as a failure to let market interests arbitrage fruit prices with access to industrial operations elsewhere in New Zealand and overseas, or as a rare victory for less profitable household and

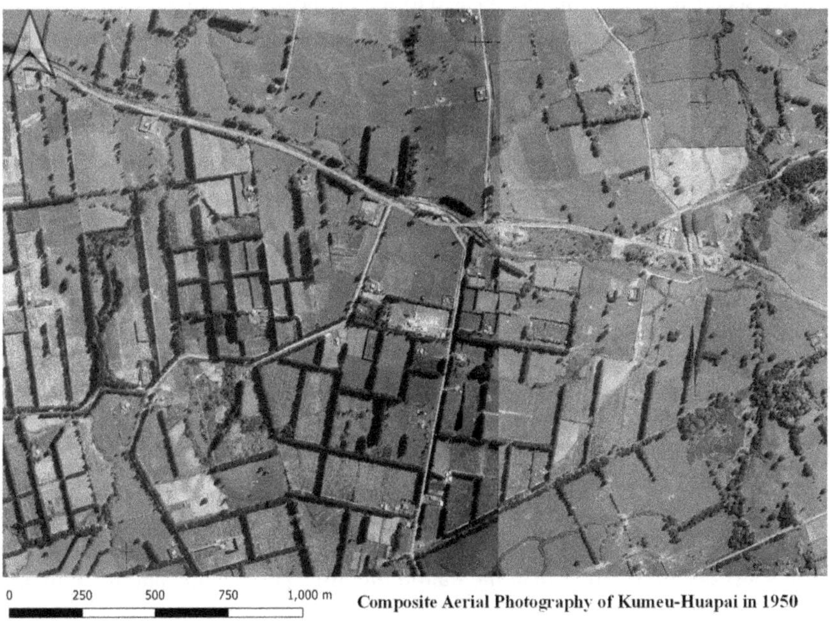


*Figure 2.7* Composite of aerial photographs collated by the LGGA’s Retrolens project (Retrolens – Historical Imagery Resource, 2020) taken of Kumeu-Huapai in 1950. Note the cluster of smaller horticultural sections south of the main road, the livestock pasture north of it, and the very small number of dwellings in the town centre.

community-level production over financial ambitions, none could deny that possibilities existed for smaller producers that can longer be imagined today.

The advantages of peri-urban horticulture as well as livestock agriculture in close proximity to inputs, markets, and labour from the city made places like Kumeu-Huapai more productive per hectare than much larger operations in places such as Hawkes Bay and Otago (Moran, 1980, p. 234), even if the former could never compete against the latter in a deregulated market with cheap international transport. Just as the flurry of development and construction work across Auckland's urban periphery today brings with it a myriad of attendant industries, Kumeu-Huapai's few decades as an important agricultural centre led to a corresponding commercial and social expansion. From being little more than a general store, train station, stockyards, and service stop in 1940, Kumeu-Huapai's town centre grew rapidly over the 60s and 70s (Dunsford, 2002, p. 214) (see Figures 2.8 and 2.9). Increased access to cars and the development of paved motorways from Auckland outwards to peri-urban rural towns like Kumeu-Huapai and further on into neighbouring regions encouraged commercial activity, even as it accelerated the process of linking rural towns with the social and economic life of the city.

Before the deregulation of production, local historians and horticulturalists told me, the subsidized rail services held a total monopoly on any freight moved more than 150 kilometres (making newly viable long-distance trucking services another sector of public life that wished to see the post-war management structure dissolve), and before automobiles were commonly accessible and the motorways were built, social life was necessarily restricted mostly to "a 20 kilometre radius for everyone except notables, you know, a doctor, a lawyer", but both of these defining characteristics of early/mid-century work and life in Kumeu-Huapai would be changed by the dominance of the car. Once Kumeu-Huapai became easily reached by car in the 1960s, many more of the aforementioned "white collar orchardists" were able to move from the city to try their hand at small-scale peri-urban horticulture, and it became ever easier for small operations to drive produce, inputs, and labour to and from the city. In this way, the success of peri-urban horticulture in Auckland and other cities is specifically an urban and centralizing phenomenon rather than a rural, local one. Writing on small-scale horticulture at the margins of cities in California today, Brinkley pointed out "the intensity and diversity of peri-urban agriculture is shown to be a result of pairing with urban markets, and not innate qualities of the soil" (Brinkley, 2018, p. 217), and many who were involved in horticulture in the 60s and 70s agree that it was the rail line (O'Ryan, 1992, p. 44), the short distance by car to Auckland, and the collective labour of the orchardist community that made Kumeu-Huapai "the biggest in Auckland" for those short decades, rather than the humid, flood-prone climate, and the only occasionally good soil.

At this point in a broad historical narrative of the transformation of a rural town into a suburb of a growing city, before I move from the dominance of one industry to another, it is important to describe the actual working experience of horticultural production in Kumeu-Huapai, at least as it is recorded by the likes of O'Ryan (1992) and Dunsford (2002), and as it is remembered by those current and former

*Figure 2.8* Composites of aerial photography (Retrolens – Historical Imagery Resource, 2020) showing the slow growth of Kumeu-Huapai over the 1940s.

horticulturalists who grew up in the heyday of the area as a productive centre. It was not an easy life:

> We would work for 60–70 hours a week, every day at least, for 6 months of the year, except for Good Friday … Then by April-May we alternate weekends, work 12 days have 2 off … we did a phenomenal amount of work, we sweated all day, we couldn't keep on fat, we ate like horses…

One grower remembered of the work he and his brother did for the family orchard. For young people in the household especially, the work required to sustain a small

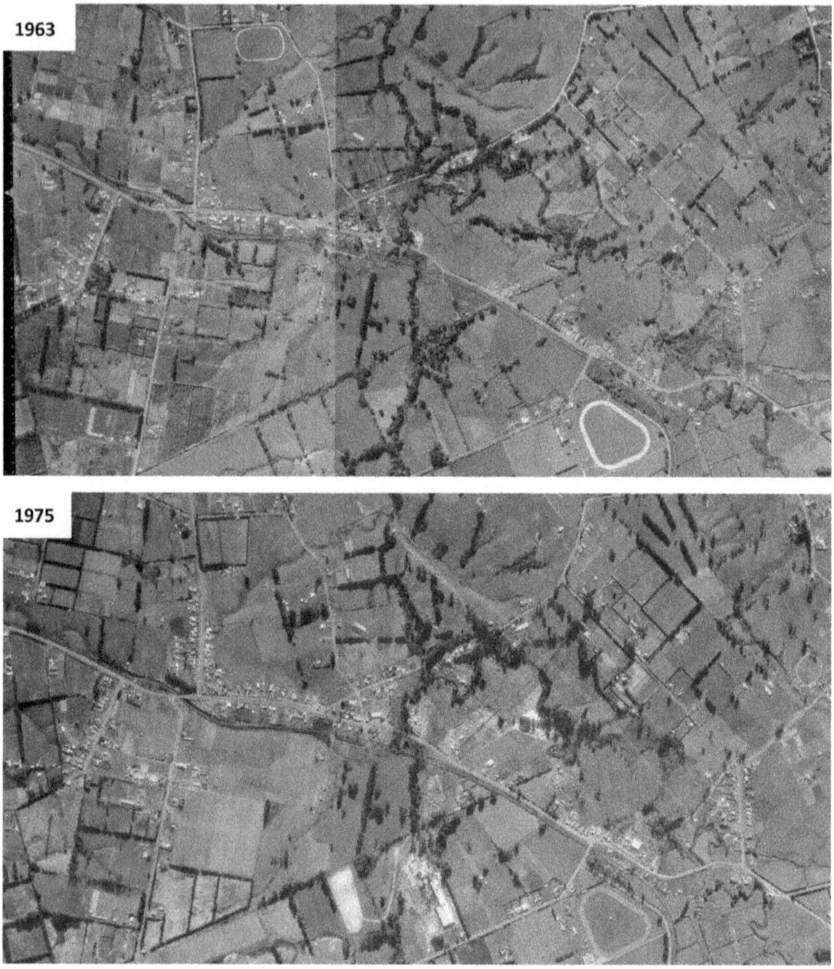


*Figure 2.9* More composites of aerial photography originally sourced from Retrolens, this time showing Kumeu-Huapai's much more dramatic development over the 1960s and early 1970s.

horticulture operation went well beyond that of any regular paid employment, at least for part of the year. Dunsford met the children and wives of apple pickers in Kumeu-Huapai who worked until 2 in the morning packing produce in a shed for transport down the rail lines (2002, p. 185), and O'Ryan found that many of the surviving orchards he surveyed in the latter days of the trade in the early 90s relied entirely on the "life cycle" of the family running it being at the point where the owner's children (or other close relatives) were teenagers who had not yet left to find work elsewhere (1992, p. 27). Whatever feelings of nostalgia exist for that life today among the people who were once these teenagers, most were glad that they no longer had to work under such demands.

One horticulturalist, suddenly animated by the absurdities of their childhood as they stood in a packing shed trying to explain decades of their work to a young academic, started to laugh and say:

> back in the good old days we'd spray, no mask no shirt, and if I came back from the fields and wasn't dyed blue by the spray, I'd get told "If you're not blue, the trees aren't blue, go back out there and keep spraying!"

These sprays that his parents demand he be dyed blue by were extremely toxic. Another former grower remembered a man who sprayed trees with his shirt off and was rapidly killed when his pores opened as he sweated in the sun, absorbing the spray. Women in greenhouses miscarried or gave birth to deformed babies, and boys in fields became permanently sick from exposure to fumes. Even today, I was told that a land developer, in turning over the earth of a former orchard in the process of making it ready to build houses on, had released the toxic chemicals that had seeped into the soil half a century before, creating a crisis on the work site and requiring a costly hazardous waste disposal process. Horticulturalists, both current and former, are quick to follow up such dramatic stories with an acknowledgement that such pesticides have long been superseded by less dangerous alternatives or outright banned, but it is a critical part of the memories of this phase of Kumeu-Huapai's relationship to state and market interests that working life was difficult and dangerous even at its most successful and most supported. For those professionals and landowners I spoke to who had once worked in horticulture but then successfully moved into a more lucrative or less physically demanding form of employment as New Zealand's economy was liberalized, the idea that people would choose to work in such a life if an alternative was offered was, by their own life experience, an impossibility.

## A return to the rule of finance

By the early 1980s, tendencies that had existed in Kumeu-Huapai for more than a century but had been at least partially countered by state-led modes of production and circulation for some decades began to make their presence known. As "the biggest in Auckland", some of the orchardists in the river valley were at the forefront of attempts to deregulate the industry in New Zealand, thinking they could expand and profit from an end to subsidized production and centralized circulation in horticulture, even as evidence from the time indicated they each benefitted greatly from the protections offered by the boards (Dunsford, 2002, p. 191). By 1983, subsidized rail transport for produce began to be phased out, dealing a significant blow to the viability of a local operation supplying fruit and vegetables for national or export sale. The privatizations of the boards proceeded as described in the previous chapter, and while many of the individual boards continued to exist for some time after the 80s, the end of a protectionist posture in state and industry interests would see a corresponding decline in the benefits for smaller operations. "The first to go is the packhouse, costs are too high" one horticulturalist explained, going to say there were now no packing houses (for commercial scale at least, as I saw a few small packing houses attached to orchards for retail purposes) in Auckland, forcing the remaining local growers to send their produce on long trips to industrial-scale centralized packing operations.

As the difficulties mounted for growers, a slow corresponding change in land use policy took place. The buying and selling of land never truly stopped being an important business. One orchardist, Tom Spraggs, since remembered as "10 acre Tom" (Dunsford, 2002, p. 240) sold his operation as Kumeu-Huapai grew and founded the Kumeu Land Agency back in 1967. He went on to specialize in the subdivision of larger rural blocks into the 10-acre sections favoured by small horticulture operations and city residents seeking to purchase a rural life. Government policy put a hold on further subdivisions in the area in 1970, seeking to preserve growing land in the Auckland region from the burgeoning rural towns that made their money from the very land they began to encroach on (p. 240). Yet, this changed with the first of several sudden booms and busts in horticulture over the second half of the 1980s, caused by exposure to international produce markets. The sudden loss of viability led state policy to first allow subdivision so that smaller orchards could be sold to a new influx of eager "white collar orchardists" hoping to catch what had become a lucrative financial opportunity and then to subdivide these orchards further to 1 or 2 hectares (in exchange for cultivating restoring some other block of land for native bush) in the face of the local commercial collapses that immediately followed (p. 252). By the time of O'Ryan's research in Auckland's northwest in the first years of the 1990s, horticulturalists were broadly pessimistic or downright fatalistic about the future of their own work (O'Ryan, 1992, p. 23), and as discussed in the previous chapter, they saw the rising speculative value of their own land as one of the few ways to either keep some part of their operation going to escape the business entirely (p. 179). Whether remembered as an annexation of territory to a new financial authority or as an end to needlessly constraining government intervention in favour of the market, food production at Auckland's urban periphery now had to compete with more lucrative land uses, and rural land at the edge of a growing city had unmatched speculative financial potential.

Many landowning residents I spoke to who had not lived in Kumeu-Huapai for their entire lives had in fact arrived and purchased lifestyle blocks in the 90s and 2000s as this collapse and reorganization were ongoing. The previous chapter mentioned that the Kumeu Show withered as an agricultural trade show in the 90s before a resurrection as a more family-focussed fair over the new millennium, and social life in Kumeu-Huapai followed a similar trajectory. Just as the Huapai Estate a century ago was rather misleadingly advertised to Auckland residents as an escape from the alienation and drudgery of urban life to a bucolic paradise, the small "lifestyle living" blocks subdivided from failed livestock farms and orchards were sold to city-based professionals (my parents included, though they moved to an adjacent area rather than Kumeu-Huapai itself), who sought to find seclusion and a rural aesthetic a short commute from their places or working and social life. Also, like the Huapai Estate, the reality of peri-urban lifestyle block living can struggle to meet the expectations formed by the real estate agent and the prospective customer. "The great thing about lifestyle blocks is you get to sell them more than once" one agent remarked, indicating that many who moved to Kumeu-Huapai over his decades-long career, like many urban professionals moving to other frontiers of Auckland (Curran-Cournane et al., 2016), were surprised and disappointed by the realities of rural production that continues around them in their new countryside retreat. Residents looking back on their first years living in the area often held a romantic image of the area in the 1990s and 2000s before development began to accelerate, "when I moved here this was a lovely rural town", "this was THE rural stop", and "Lifestyle block paradise" being three particularly memorable descriptions of the river valley. Horticulturalists, both current and former, remember the new resident's love of the land differently. "We get that rurally, from people who moved there for the peace and quiet … people having a BBQ asking why you're working on a Sunday ruining the afternoon" one grower told me, before launching into a tale about a new neighbour of theirs attempting to kill off all of the insects and plants around their country home with the zealous use of extremely powerful pesticides, something the horticulturalist found deeply ironic given they had just been regaling me with stories of "the good old days" of mid-century horticulture and its cavalier approach to toxic chemicals.

These two particular complaints – the productive work of horticulture interfering with the enjoyment of leisure and the encroachment of the local ecosystem onto the built environment – did much to accelerate Kumeu-Huapai's transition from a rural town to a suburb-in-waiting of Auckland City by encouraging further regulation of horticultural activities. As in Pukekohe south of Auckland today (Curran-Cournane et al., 2016), remaining horticulture operations began to be restricted in their operating hours and techniques for the sake of their new neighbours, a change that unavoidably "makes things non-feasible" as one former grower said. It is this final change in resident's relationship to land, one dedicated to suburban rather than productive land uses, that proves to be the most irreversibly transformative. "They're already around us, it's like at a battle, where some battalion leaves a hole in the line, lets the enemy through and surrounds us" one of the two remaining operators explained, going on to repeat a line they had earlier said over the phone to me "the horses have bolted, it's long gone".

## Water and sewerage: public enablement of private suburban development

From the 1990s onwards, the attractiveness of the area for urban residents seeking lifestyle block living only increased. The population of the area saw nothing like the boom that defined the 2010s (see Figure 2.10), but the transformation of social and working life for the residents of Kumeu-Huapai continued apace through the rapid conversion of productive land into lifestyle living across the northwest of Auckland. Under this initial phase of urban expansion, houses were built (or moved from elsewhere) on newly subdivided land for professionals seeking a slice of rural life, arriving in limited but steady numbers as farmers and landowners sold off and moved on. The true catalyst for the acceleration of development and the rapid suburbanization of Kumeu-Huapai was, I was told by a couple of real estate agents and developers and a few residents, the provision of centralized water and sewerage. Those who, like my parents and several of my interlocutors, had purchased a relatively cheap section cut from a failed farm or orchard in the urban periphery some years before Kumeu-Huapai's expansion began in earnest drew their water from rain collection or groundwater bores and installed septic tanks for waste. Many lifestyle blocks in the area continue to operate on an independent system, but at the turn of the millennium, reticulated water and sewerage infrastructure began to link Kumeu-Huapai with Auckland's central system. Cabra bought much of the land around Tapu Road at this time (long before the current managing director I spoke to had joined the company), spurred into action by the knowledge that a greater scale of urban development would soon be feasible. With the capacity of the area to support rapid urbanization supported by new water and sewerage infrastructure, the first spate of development around Tapu Road and the creation of the Huapai Triangle, the first and largest of the Special Housing Areas on the land that had once been Nobilo's winery, became possible.

Yet, those who seek to make their fortunes capitalizing the land of the Kumeu River Valley anew today are faced with a series of related problems. The rapid urban expansion makes the area harder to sell on the merit of rural isolation and bucolic bliss, leading some existing residents to seek a new lifestyle block further from Auckland's urban centre (and make a massive profit from the appreciation of their land values as they do so) and potential buyers to reconsider any notions given to them about a rural life in reach of the city. Developers who turn to suburban or even dense urban forms of development at Auckland's urban frontier in the face of the loss of a marketable "rural" quality to land then find that water and sewerage continue to be the prime limiting factor. The managing director of Cabra and

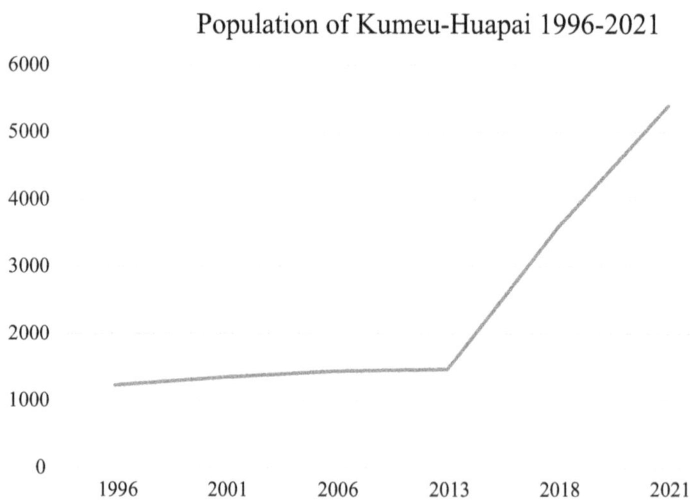


*Figure 2.10* The population of Kumeu-Huapai from Stats NZ, drawn from 1996 to 2018 Census data and projections of population change in between this last census and 2021 (Stats NZ, 2022). Note the slow increase followed by the massive rise in population in the last seven years, as mentioned in the previous chapter.

the CEO of Fletcher Residential each acknowledged that this physical limit on the amount of (current) development capacity made their projects intensely competitive and necessarily as rapid as possible, "the current system promotes a first in best dressed" the latter remarked. This pressure to be the first to transform rural land into a brand-new suburb before new developments cannot be supported is felt by local government in the area and in Auckland City, "if Fletcher's gets to leapfrog it ... It doesn't magically give more money for community infrastructure" one official complained to me. At the time of my fieldwork, this problem of water and sewerage capacity was a hard limit on further ambitions to suburbanize Auckland's northwestern periphery (Figures 2.11 and 2.12).

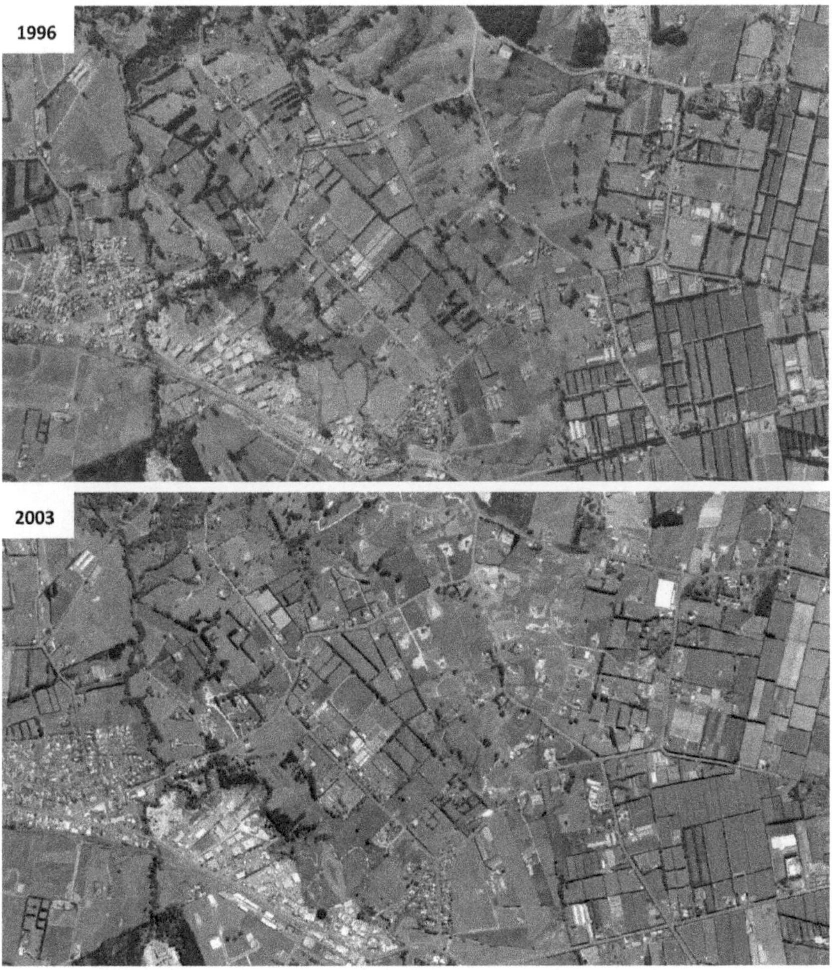


*Figure 2.11* Steady development of small lifestyle blocks around northeast Kumeu-Huapai from the 1990s onwards. Aerial imagery sourced from Auckland Council GeoMaps (Auckland Council, 2016).

*Figure 2.12* The same part of northeast Kumeu-Huapai between 2010 and 2015. Note how the development of small lifestyle blocks is now being matched by more traditionally "suburban" patterns of urban development.

In order to continue the process of more intensely capitalizing the land of the river valley, vast amounts of public money will have to be spent on making greater urban capacity possible, and the largest piece of wastewater infrastructure in the entire country must be built. A project currently referred to as the Northern Interceptor will divert wastewater from Auckland's northwestern urban frontier out to the North Shore, and along with the NH2 water mains project, will provide that basic support for water and sewerage for the new suburbs that have yet to be dug and sculpted from the hills and flood plains around Kumeu-Huapai (Watercare, 2020). Developers publicly and enthusiastically support new efforts to extend public transport networks to make living in the area more feasible for urban professionals, the CEO of Fletcher Residential telling me "There's no logical constraint" for getting train services out west on the same line that once carried produce grown in Kumeu-Huapai to the city and across Aotearoa in different times. They also call for new community facilities, the managing director of Cabra saying "they (the government) should be building a secondary school now" before expressing irritation at the slow pace of local and central government response to their needs.

The counter from the public sector is simple: "there's just no money … Watercare is totally stretched … Auckland Transport has eight billion dollars in unfinished projects". For capitalists operating in Kumeu-Huapai today, the next phase of making the land available to be capitalized depends on the state being not only aligned with their interests but able to marshal a massive amount of resources to enable forms of development that suit them. The intentional privatization and enclosure of economic life that has come to define the past half-century of ruling policy across the planet (Bourdieu, 1998) has left local and central government simply too stripped of income and power to reorganize the land and people of the Kumeu River Valley in such a way as to open the doors for more profit-making. For this reason, real estate agents, developers, and experts of all stripes today are waiting to see what happens and making what noise they can about the importance of infrastructure and community facilities, before they commit to more development, at least openly. Yet, as mentioned in the previous chapter, Fletcher Residential has recently acquired two large plots of rural land, one in a flood plain south of Kumeu-Huapai and the other in an existing agricultural operation northeast of it. Should they have the necessary connections – and many assure me that no other land development company operating in Aotearoa can match them in that respect – they may seek to take the remaining development capacity now, before a competitor can do the same.

## Selling the Fringe: historical tendencies of the capitalization of the Kumeu River Valley

Through more than one-and-a-half centuries of efforts to render the land of the Kumeu River Valley into a source of profit, some common historical tendencies – and corresponding interested parties – can be discerned. These tendencies have been present from the initial acts of the dispossession and privatization of land tenure from Ngāti Whātua into the hands of speculators backed by the British Empire, to forestry and early attempts at agriculture by those few non-speculative buyers, to the reinvention of the area as the Huapai Estate to be sold to hopeful colonists and city dwellers, and to the decades of bitter work and state-protected production that made the valley into a horticultural and agricultural centre, all the way to the reassertion of speculative financial demands and the accelerating urbanization of the entire region. The first and clearest is that the lands of the river valley have been most profitable to those who used it as a financial product rather than as a place to live, to rest on a journey, or to grow food. Just as the elder William Deloitte quickly sold his acquisition of Hokianga's port to an Australian enterprise and moved on to other financial endeavours, leaving the material work of colonizing the land to those who came after, the junior Deloitte concerned himself only with the acquisition of the rich farming land of the Taupaki block at as low a price as possible and the arrangement of mortgage agreements for the Waikoukou block. Many of the people who bought land from Deloitte in turn also did so as an entirely speculative venture, and it was decades before Taupaki was occupied by more than a few families (Dunsford, 2002, p. 26). The many hundreds of acres acquired by Thomas Deacon through financial dispossession against Ngāti Whātua were treated in a likewise manner. The extraction of value from land and labour – cutting down the rainforest, digging for Kauri gum, initial attempts to cultivate livestock and crops – were all physically demanding and often not very profitable ways to make a living, and thus better left to the peasants and urban professionals who came to the frontier of the empire to seek a change of fortunes.

This pattern repeats with the creation of the Huapai Estate by Auckland shareholders and its advertisement as a rural paradise to unwitting buyers, followed by the selling of Kumeu-Huapai as a "lifestyle block paradise" and then a suburban compromise between urban social life and rural aesthetic in recent decades. One resident neatly encapsulated this dominant attitude to land in the river valley in their complaint of the development of the Huapai Triangle: "Landowners got millions in untaxed gain, everyone else got left with the work". Every other purpose for Kumeu-Huapai's land, be it productive or recreational, has existed in relation to its speculative financial value.

Secondly, every great financial success for any landowner can only occur with the backing of the state or at least the alignment of state and market interests. For private buyers or creditors to legally expropriate land from Tangata Whenua, the centralized process of crown purchases to finance colonial development had to be discredited and deregulated, while the land-holding rights of these private capitalists had to be protected by the same colonial government as the rights of tribal or collective tenure were ruled out. The pressures of the Great Depression and the Second World War demanded that state interests focus on centralized production and circulation over other concerns, and this necessity helped lead to the mid-century success of peri-urban agriculture that residents today remember as the real expression of bucolic colonial origin once falsely promised by the creators of the Huapai Estate. The building of motorways and the state highways by the very same central government bought Auckland and its rural frontier ever closer together, allowing both the further intensification of local horticulture and the beginnings of an exurban lifestyle for urban professionals. The dramatic realignment of the state to the desires of the market in the final quarter of the 20th century ensured that treating peri-urban growing land as a financial product or as an aesthetic value – both tendencies that had been strong even when the state had organized Kumeu-Huapai as a place to grow food – became the overriding factors for the area once again. Today, the success of the next group of landowners, financiers, and developers at Auckland's northwest frontier depends on the capacity of the Auckland Council and the central New Zealand government to complete a series of massive infrastructure projects that will enable and justify further suburban sprawl. Whether this is to be sought through the personal relationships detailed in the previous chapter or through overt public petitioning, land developers and financiers need state intervention to see their plans to capitalize the land of the river valley come to fruition.

The third crucially important historical tendency that has defined the development of Kumeu-Huapai is the presence of paying customers and prospective mortgagees who provide the demand for land at the periphery of Auckland. Deloitte and Deacon needed buyers to realize the speculative value of their initial primitive accumulation, the Huapai Estate could not have been a successful venture had their advertising not successfully drawn in people with the inclination to escape urban life and the means to purchase the land, and developers today rely on the existence of a large number of urban professionals (mostly couples who both work full time) with the salary to cover a > $1.3M NZD mortgage in order to fill the demand for their maximally profitable model of suburban development. More broadly, the presence of Auckland's concentration of wealth, labour, and expertise also made possible the success of small-scale peri-urban horticulture, so long as state interests were concerned with preserving such a land use in areas close to the growing city (Brinkley, 2018; Moran, 1980). Seasonal workers hired from Auckland (during the Great Depression many of them lived in temporary work camps in the area), alongside the urban professionals and their relatives who had so optimistically aspired to live in a colonial paradise by buying a section in the Kumeu River Valley, were the ones whose decades of work transformed a deforested swamp into a regional centre of horticulture and a thriving rural town. The social life of Kumeu-Huapai residents has become more explicitly linked to city life as these same city dwellers became more able to live in the urban periphery and commute to work as roads and vehicles improved, but the history of the river valley has been as much in relation to the life of the city as it has with the ambitions of finance and the juridical forms of the colonial state.

These tendencies and interests in colonial Kumeu – the individuals and organizations that speculate and financialize the future values of large swathes of land to great private profit, the relationship of the state to working life and land uses, and the desire of urban professionals to seek an escape in rural life – have been present in the river valley since capitalists first sought to privatize the purchasing process of land from Tangata Whenua. Yet, these "coexisting variations" are, Ferguson put it, "continually modified in complex and nonlinear ways" (Ferguson, 1999, p. 42) over the subsequent 150 years of Kumeu-Huapai's formative history, rather than proceeding in a single direction from some received image of rugged colonial frontier naturally towards the culmination of its destiny as the newest northwestern suburb of Auckland City. If, as Derry proposed, the development of Kumeu-Huapai is a form of "growing up", then the Kumeu River Valley has had a difficult and circuitous adolescence with more than one false jump towards a supposedly inevitable suburban maturity, and developers today seem worried that the formerly rural town may never fully grow up without an intervention from the government. The particularly pressing global historical contingency of a massive economic collapse followed by a half-decade war effort created the conditions for small-scale horticulture's best days in Kumeu-Huapai, and the reaction of capitalists and business owners against restrictions and regulations some four decades later was equally a result of global and national shifts in the relationship between state and production as much as it was a project of some successful landowners in the area. The urbanization of horticultural centres closer to Auckland over the past century provides a powerful narrative of an endless expansion of urban influence, but much had to be done by interested parties to ensure that the same story began to take shape in Kumeu-Huapai.

The very close relationship between the land uses considered best for the Kumeu River Valley and the desires of financial interests to achieve a profitable sale on speculations of future value is as old as the expropriation of the area by economic and juridical means in the 1860s and has never entirely been absent since. Yet, depending on the forms of capitalization possible and the relationships between state and market interests, this financializing tendency that is so visibly dominant in Auckland's urban periphery today has been less so in the past, and it has also been expressed in different forms due to the limitations of the situation any individual capitalist in Kumeu-Huapai found themselves in. While perhaps a subtle distinction from a broad narrative of unidirectional linear change towards an inevitable strip-mall future, this understanding of Auckland's urban development allows us to consider how it was decided that Kumeu-Huapai would surely be best suited as a future suburb of the growing city. To this end, the following chapter will focus on the ways land in Kumeu-Huapai has been valued in the past and today, and how state and market interests have vied to shape depictions of the landscape to serve their interests.

## Notes

1 These towns are regional centres, with populations of between 50,000 and 80,000 people.

2 Iwi that have received some form of settlement from the New Zealand government as an initial reparation for colonial expropriation set up trusts and corporations in order to manage the assets that have been handed. The development trust for Ngāti Whātua O Kaipara today manages some local forestry blocks, recreational reserves, and the land on which local public schools sit.

3 The resin of Kauri trees that once covered the North Island in vast forests, Kauri gum can be dug from the earth where the trees once stood and was used as a varnish, a fire-starter, and in early iterations of linoleum (Walrond, 2007).

**Bibliography**

Auckland Council. (2016). *Auckland Council GeoMaps*. https://geomapspublic.aucklandcouncil.govt.nz/viewer/index.html

Ballara, A. (2003). *Taua: 'musket wars', 'land wars' or tikanga? Warfare in Māori society in the early nineteenth century*. New York: Penguin Group.

Bourdieu, P. (1998). The essence of neoliberalism. *Le Monde diplomatique*. Retrieved December from https://mondediplo.com/1998/12/08bourdieu

Bozic-Vrbancic, S. (2006). Mysterious degrees of whiteness: Stereotypes about Croats in colonial New Zealand. *Revija za sociologiju*, *37*(3–4), 181–192.

Brinkley, C. (2018). High rugosity cities: The geographic, economic and regulatory pathology of America's most non-concentric urban areas. *Land Use Policy*, *73*, 215–224.

Carter, M. (2016). Preliminary report on the Archaeological Excavation of the New Deptford Shipbuilding Yard (O05/317), Horeke, New Zealand. *Archaeology in New Zealand*, *59*(2), 24–38.

Curran-Cournane, F., Cain, T., Greenhalgh, S., & Samarsinghe, O. (2016). Attitudes of a farming community towards urban growth and rural fragmentation—An Auckland case study. *Land Use Policy 58*, 241–250.

Dunsford, D. (2002). *Doing it themselves: The story of Kumeu, Huapai and Taupaki*. Auckland: Kumeu District History Project.

Ferguson, J. (1999). *Expectations of modernity: Myths and meanings of urban life on the zambian copperbelt*. Oakland: University of California Press.

Google LLC. (2022). *Google Satellite via QGIS* [XYZ Tile Data showing Satellite Imagery of Western Auckland]. Retrieved from http://mt0.google.com/vt/lyrs%3Ds%26hl%23Den%26x%23D%27Bx%27D%26y%23D%27By%27D%26z%23D%27Bz%27D&zmax=18&zmin=20

Graeber, D. (2011). *Debt: The first five thousand years*. New York: Melville House.

Harvey, D. (2017). The 'new' imperialism: Accumulation by dispossession. In B. Ollman & K. B Anderson (Eds.), *Karl Marx* (pp. 213–237). London: Routledge.

Moran, W. (1980). Mechanisms of land use allocation on the urban periphery, an Auckland case. In A.G. Anderson (Ed.), *The Land Our Future* (pp. 223–238). Auckland: Longman Paul.

New Zealand Herald. (1887). *Death of Mr. C. O. Davis*, p. 6. Retrieved June 29 from https://paperspast.natlib.govt.nz/newspapers/NZH18870629.2.56

O'Ryan, M. J. (1992). *Orcharding in the urban periphery: An examination of family orchard management strategies in Auckland's Northwestern Periphery*. Auckland: The University of Auckland.

Parker, R. H. (1984). *Papers on accounting history*. New York: Garland Publishing.

Retrolens – Historical Imagery Resource. (2020). Retrieved June 1 from http://retrolens.nz/

Ryburn, W. R. (1999). *Tall spars, steamers & gum: A history of the Kaipara from early European settlement, 1854–1947*. Kaipara Publications.

Stats NZ. (2022). *NZ.Stat*. Retrieved May 18 from https://nzdotstat.stats.govt.nz/wbos/index.aspx

Stirling, B. (1998). Ngati Whatua Native Land Court block histories, 1864–1900: The evidence. Auckland: Crown Forestry Rental Trust and Ngāti Whātua o Kaipara Ki Te Tonga Claims Committee.

Walrond, C. (2007). 'Kauri gum and gum digging'. Wellington: Te Ara – The Encyclopedia of New Zealand.

Ward, A. (1990). 'Davis, Charles Oliver Bond'. *Dictionary of New Zealand Biography*. Wellington: Research & Publishing Group at Manatū Taonga Ministry for Culture and Heritage. Retrieved from https://teara.govt.nz/en/biographies/1d3/davis-charles-oliver-bond

Watercare. (2020). *Projects around Auckland – Northern Interceptor*. Retrieved May 16 from https://www.watercare.co.nz/About-us/Projects-around-Auckland/Northern-Interceptor

# 3 The axiologies of Kumeu-Huapai's land

*Figure 3.1* Land in Kumeu-Huapai's "Future Urban Zone". Photo taken by the author.

As I spoke to all of those experts in the land that surrounded the process of development in Kumeu-Huapai – horticulturalists current and former, land developers, subdivision consultants, planners, and lawyers – a common point of disagreement among them was what the land of the Kumeu River Valley was actually good for. More interestingly, positions on this matter did not always align with the interlocutor's economic interests. A sales manager for a construction company specializing in large suburban builds remarks that the land here is far better suited to growing food on than building houses on, "the soil is fertile, so the land is good for growing, not so much for construction", while a horticulturalist tells me that if they had their time

DOI: 10.4324/9781003400325-5

again, they would have bought land in central Otago (a far dryer climate than found in the Auckland region) rather than Kumeu-Huapai, "In Otago all you need is irrigation, but up here you need fungicides … new fruit variations are hard to grow here".

Attempts to quantitatively assess the true "value" of Kumeu-Huapai as a horticultural centre are hamstrung by the diversity of soil types across a single 10 acre/4 ha block, let alone the extent of Kumeu-Huapai itself. Unable to make any clear categorizations for the entire river valley, soil scientists and surveyors must decide how they can most accurately represent and classify the land in a way that is legible at a regional scale while still being representative of the material realities of the soil at any one point across Kumeu-Huapai. The result is that classifications of land use in the river valley over the past half century have varied significantly in detail even if general similarities are consistent across mapping efforts and that correspondingly large variations are made about the best use of land here. In this environment, where the particularities of the soil of any two particular orchards can vary greatly despite being classified similarly, there is plenty of room for an interested party to claim a particular provenance for the land and for well-informed professionals to disagree on a categorization of said land based on their own sources and interpretations. This chapter examines the necessarily interpretive and creative process of shaping the axiologies (systems and analyses of value) of rural land in Kumeu-Huapai, with a particular focus on how land becomes defined as either notably valuable for horticulture or as difficult soil best suited for suburban development. How such a distinction is decided provides an insight into a less-visible yet crucial part of the process of urban development – what land is allowed to be used for and why.

As the historical narrative of the previous chapter alluded to, the actual biophysical properties of the soil found in Kumeu-Huapai are neither the only nor even the most important factor in the viability of small-scale peri-urban agriculture, and the soil itself is not an unchangeable "natural" constant. What was once a rainforest was left a swamp by the first round of extractive labour at the end of the 19th century. This was followed by a second transformation into an agricultural and horticultural centre of the region through decades of draining, fertilizing, and maintaining the earth. Most recently, a third dramatic change in the physical makeup of the river valley brought about by horticultural collapse and an explosion of suburban development in the last quarter of the 20th century. The close proximity of the railway line, built by Māori and Irish labourers on the land gifted by Ngāti Whātua in exchange for a quickly broken promise of free passage on the line for the iwi (Dunsford, 2002, pp. 31–32), radically altered the equation for viable land uses in the area by ensuring a freight route to Auckland and the rest of the country. The construction of state highways and the greater availability of cars further compressed the distances between the urban economy and Kumeu-Huapai's land for the purposes of both production and the selling of a rural aesthetic to city dwellers. Each of these material and social transformations redefined what Kumeu-Huapai's land could be valued for.

The land of Kumeu-Huapai, and more broadly the landscape of the river valley itself, is "a concretization or reification of the social relations that go into its making", as Don Mitchell put it in his own study of the land near where he grew up in California (2003, p. 10). Chapter 1 sketched out the relationships that define the rise of property development and concommital collapse of peri-urban horticulture in Kumeu-Huapai, and Chapter 2 provided historical context for these relations and interests, and thus this chapter will demonstrate how these interests have determined what land in the Kumeu River Valley should be used for today. I will present past and present attempts to classify land uses in the area and relate them to the changes in social and economic relationships – centralized production to privatized and industrialized horticulture, the rise in exurban lifestyle living, and urban sprawl – to show how the values held towards peri-urban land have been changed in order to present a new reality for the area. Mitchell's interpretation of Marxist geographies of landscape provides an incisive lens through which to view such change:

> Ontologically, landscape is best conceptualized first (though not exclusively) as dead labor. That is to say, the landscape is a social product, it is something made. And it is constantly remade. But to say the landscape is constantly remade is not to imply that it is always and fully in a state of flux. On the contrary, the landscape must also be understood as a "structured permanence".
> (2003, p. 16)

The "structured permanence" of Kumeu-Huapai's landscape today is a drastically different one than the place that residents knew and saw as equally natural four decades ago, and the axiologies of land in these two time periods reflect such a gulf in perceived realities.

## Class conflicts: land use classification in Auckland now and half a century ago

The suitability of land for productive use in Aotearoa today is classified by environmental scientists in a hierarchical system of classes from 1 to 8 (in descending order of productivity) with additional modifiers to provide details on dominant mitigating factors (LRIS Informatics Team, 2011). This system is referred to as Land Use Capability (LUC). For example, if an area of land is classified as 2W, this means it has been determined that the land has a very high productive potential, only held back from being even more valuable for its poor drainage or high water table (W literally standing for Water), while a section marked as 7E is very poor due to severe erosion. See Tables 3.1 and 3.2 for a detailed breakdown of these categories.

This classification system is an adaption of one created by the US Department of Agriculture, and it was adopted by the NZ government in the middle of the 20th century. In the decades since, the system has necessarily been applied and interpreted on a regional or local basis, at whatever extent has been feasible for the surveying team. For this reason, identical soil formations have different LUC codes depending on whether they are located north or south of Auckland City (Hicks & Curran-Cournane, 2017, p. 3), and I would sometime hear interlocutors wave away an area of land as "class 1, but not really that good" or tell me that a section was excellent for cultivation precisely because of a combination of qualities (ruggedness of terrain and stoniness of soil) that usually hindered horticulture. Despite these

*Table 3.1* The eight hierarchical classes of land use classification

| *LUC Class code* | *Description* |
|---|---|
| 1 | Land with virtually no limitations for arable use and suitable for cultivated crops, pasture, or forestry |
| 2 | Land with slight limitations for arable use and suitable for cultivated crops, pasture, or forestry |
| 3 | Land with moderate limitations for arable use, but suitable for cultivated crops, pasture, or forestry |
| 4 | Land with moderate limitations for arable use, but suitable for occasional cropping, pasture, or forestry |
| 5 | High-producing land unsuitable for arable use, but only slight limitations for pastoral or forestry use |
| 6 | Non-arable land with moderate limitations for use under perennial vegetation such as pasture or forest |
| 7 | Non-arable land with severe limitations to use under perennial vegetation such as pasture or forest |
| 8 | Land with very severe to extreme limitations or hazards that make it unsuitable for cropping, pasture, or forestry |

*Source:* Taken from LRIS NZ (LRIS Informatics Team, 2011, p. 7).

*Table 3.2* The subclass modifiers indicating the dominant limiting factor of the land's productive potential

| *LUC subclass modifier* | *Description* |
|---|---|
| e | Erosion susceptibility, deposition, or the effects of past erosion damage first limit production |
| w | Soil wetness resulting from poor drainage or a high water table, or from frequent overflow from streams or coastal waters, first limits production |
| s | Soil physical or chemical properties in the rooting zone such as shallowness, stoniness, low moisture holding capacity, low fertility (which is difficult to correct), salinity, or toxicity first limit production |
| c | Climatic limitations such as coldness, frost frequency, and salt-laden onshore winds first limit production |

clashes between a formalist categorization system and substantive local understandings of land, and the fact that Land Use Classification interpretation is based on land mapping from the 1970s with a limited set of updates in the 1990s (Hicks & Curran-Cournane, 2017, p. 8; LRIS Informatics Team, 2011, p. 2), this way of delineating the relative productive values of land at Auckland's urban periphery is the most commonly and consistently applied metric in any discussion of urban sprawl.

As Auckland City expands and the unmatched profitability of residential development leads peri-urban landowners to make their case to local government for being allowed to sell their rural land to land developers (this dynamic will be explored in detail in the following chapter), LUC is one of the metrics that has been utilized in planning and legal contexts in an attempt to provide clarity to these discussions upon which large amounts of money stand to be made and permanent changes to the lived environment inflicted. Since a ruling made in New Zealand's Environmental Court in 2018, new development must be undertaken while "avoiding elite soils and avoiding where practicable prime soils which are significant for their ability to sustain food production" (The Environment Court at Auckland, 2020, p. 10). These two labels – Elite and Prime – refer to Class 1 land and to Class 2 and 3 lands with favourable growing conditions, respectively (Auckland Council, 2022, p. 66). The decision that development must avoid elite soils in general while avoiding prime soils "where practicable" if they are significant for food production does not mean that all Elite/Class 1 soils have been forbidden from being built on and that Class 2 land with a notable history of food production has remained either productive or rural, but even if this rule is regularly bent, it is the clearest line that exists for what is to be considered valuable productive land in Auckland.

If one looks at a map of the remaining Class 1 and 2 lands in the region today (see Figure 3.2), the southern town of Pukekohe, mentioned in previous chapters, retains most of Auckland's Elite or Class 1 land, even as rural production begins to collapse in the face of financialized forms of development today just as they did in Kumeu-Huapai 30 years ago. The remaining Class 2 land covers a large quantity of the surrounding southern Franklin ward of the Auckland region, and it is also concentrated around Kumeu-Huapai and the surrounding towns of Riverhead, Waimauku, and Taupaki. This distinction of value between the small-scale former horticultural productive centre and the larger (yet still local and mostly owner-operator run) extant one was sometimes an explicit part of the narratives about Kumeu-Huapai's urbanization; "if I was doing this in Pukekohe it would be a sin" one lifestyle block resident said of their decision to move from the city to the Kumeu River Valley. They explained that using the land here as an aesthetic product certainly may not be productive, but at least it was not overtly destructive of any truly valuable soil resources such as could be found in Pukekohe. Such a comparison was also present in the (as stated in the introduction of this chapter, often not entirely straightforward) justifications of urbanization by professionals involved in the process. The aforementioned factors that made the river valley difficult to grow crops in for the first settlers of the Huapai Estate in the 1920s – flooding, inconsistent soil, salt (in western areas of the estate), and uneven terrain – were all absent in the industrial-scale horticulture operations in Nelson and Hawke's Bay or the remaining peri-urban production areas in Pukekohe. So long as national and international transport networks enabled rapid delivery of produce, Kumeu-Huapai was simply outclassed by the competition, as current LUC maps would confirm.

Yet, this very clear divide in the valuation of land between vaunted Class 1 soils in Pukekohe and a few areas across Aotearoa versus Class 2 land grouped across a few distinct areas of peri-urban Auckland that can be acceptably built over so long as it can be proven to not be significant for production did not exist 50 years ago. Maps drawn up by the Auckland Regional Authority (ARA, a previous incarnation of Auckland Council) using the same land use data that underlie the Land Use Classifications seen in Figure 3.2 follow a very different logic to classifying and valuing land. A 1975 map entitled "Soil Resources – Soil Limitations for Urban Development" (see Figure 3.3) follows the eight-level hierarchical structure common to

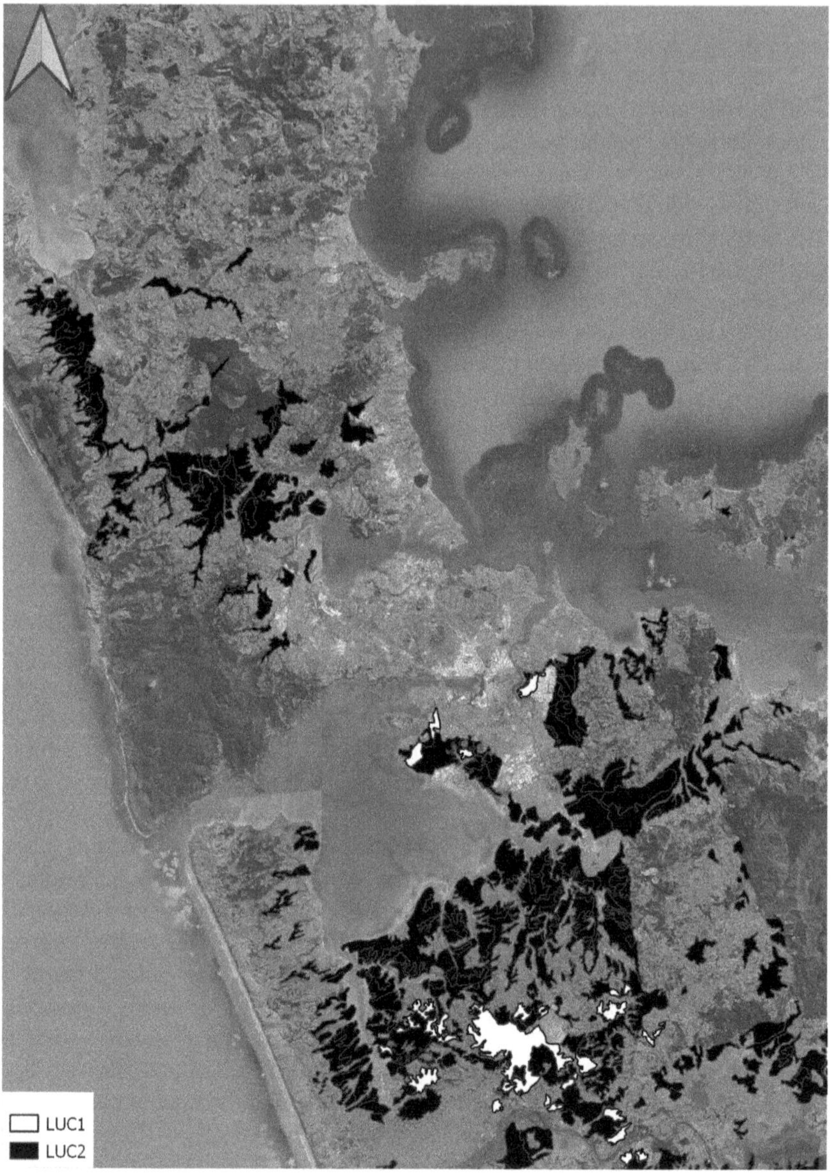


Remaining LUC Class 1 (White) and Class 2 (Black) land in the Auckland region

0 5 10 15 km

*Figure 3.2* Class 1 and Class 2 lands in the Auckland Region, data from the New Zealand Land Resource Inventory (LRIS Informatics Team, 2021). Satellite imagery from Google Maps (Google LLC, 2022).

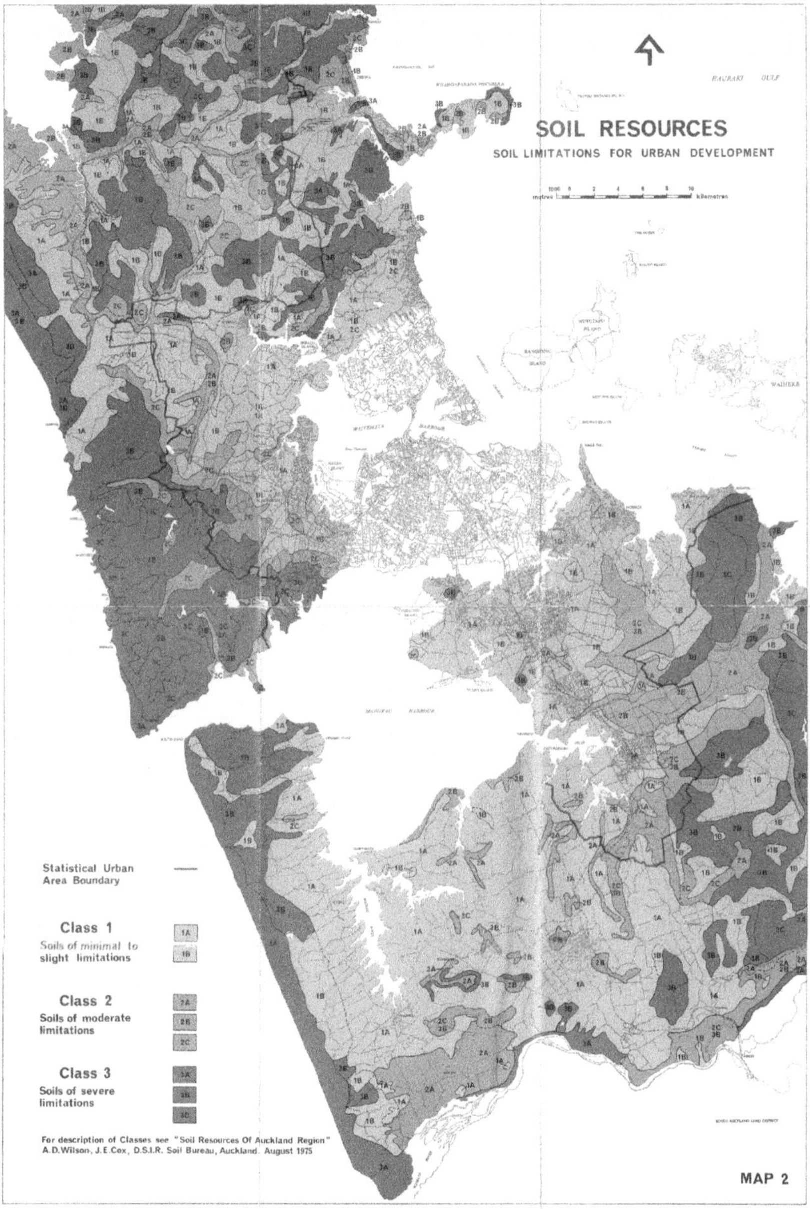


*Figure 3.3* 1975 ARA map of Land Use Classes for the Auckland Region, with the eight classes grouped into three categories of land with high, moderate, and low productive value. The hard copies of this and the following map are both possessed by Eric van Essen and are reprinted here with his permission.

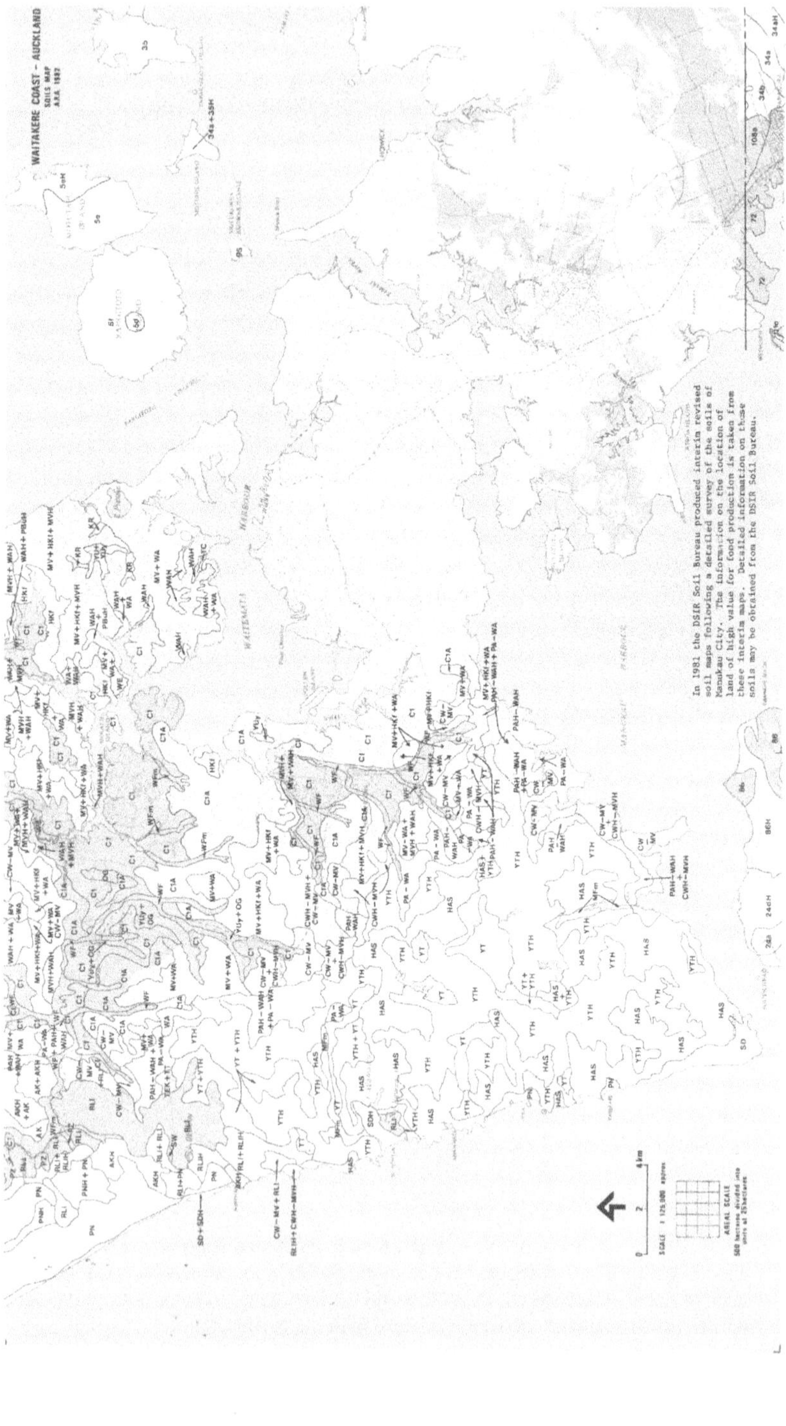


*Figure 3.4* Northwestern panel of a 1982 ARA map delineating areas of high value to food production.

the modern LUC map but applies its definitions radically differently. As in the 2018 Environment Court reading, Class 1 and 2 lands are both considered valuable and are bracketed together as 1A and 1B: Soils of Minimal to Slight Limitations. Yet, much of Kumeu-Huapai and elsewhere in the northwest is placed in the same category as Pukekohe. Northern and Western settlements urbanized long before Kumeu-Huapai such as Henderson, Albany, and Massey are similarly categorized as having the most valuable productive land in the region and the most worthy of protection from development. Another map from the ARA in 1982 aimed to more clearly delineate which land had significant productive value and which faced no such impediments to development (see Figure 3.4). Kumeu-Huapai, along with Henderson, Albany, and Whenuapai, is highlighted in bright yellow as of "high value for food production". A close-up view of Kumeu-Huapai's zoning today (including the "Future Urban Zone" for rural land soon to be rezoned for development) over these maps (Figure 3.5) demonstrates the extent to which development, both suburban and lifestyle subdivision, is conducted on productive land. Much of the Huapai Triangle and Tapu/Matua Road Cabra-led developments discussed in the first chapter sit directly on land marked as 1A on the 1975 map and highlighted as productive in the 1982 map. What changes occurred over the following half-century to downgrade northwest Auckland's productive value, along with that of urbanizing areas across the Auckland region such as Albany (north), and Beachlands (southeast) without a corresponding change in the basic data that defines Land Use Classification? As narratives in previous chapters attest, two important factors are the changing position of the state in relation to production and the growing power of the financial value of land over other considerations.

## "Not much productive agriculture going on out there now" creating and reifying the image of the peri-urban landscape as barren hinterland

Just as the fading memories of Kumeu-Huapai as a productive centre often lie at odds with the dominant presentation of the area as either a "lifestyle block paradise" or a new suburb of Auckland itself, the maps presented above are a rare insight into an ontology of peri-urban land, which is otherwise difficult to uncover today. Both the 1975 and the 1982 maps are not (to my knowledge) available online under any of the agencies that succeeded the Auckland Regional Authority or the Department of Scientific and Industrial Research (DSIR) after their various reorganizations. A soil expert noted that this diffusion of responsibility for the "seven or eight" different organizations that could be considered as having ownership of these data tended to result in it simply being lost, "everyone's responsible so in the end no-one really is". I scanned these maps from this (now retired) expert's own physical collection, and in fact, I had to ask to borrow them in turn from one of the local historians I had interviewed, who was using them for their own research into the area and did not know if any other such historical maps were readily available. Any close-up examination of how land in Kumeu-Huapai has been quantitatively valued must confront this paucity of surviving data from such a relatively recent

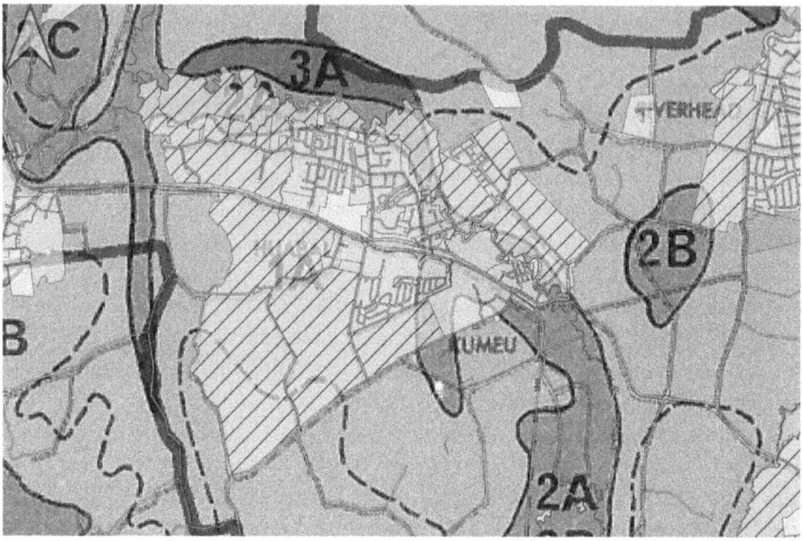


Auckland Unitary Plan for Kumeu-Huapai overlaid onto 1975 ARA Soil Resources map

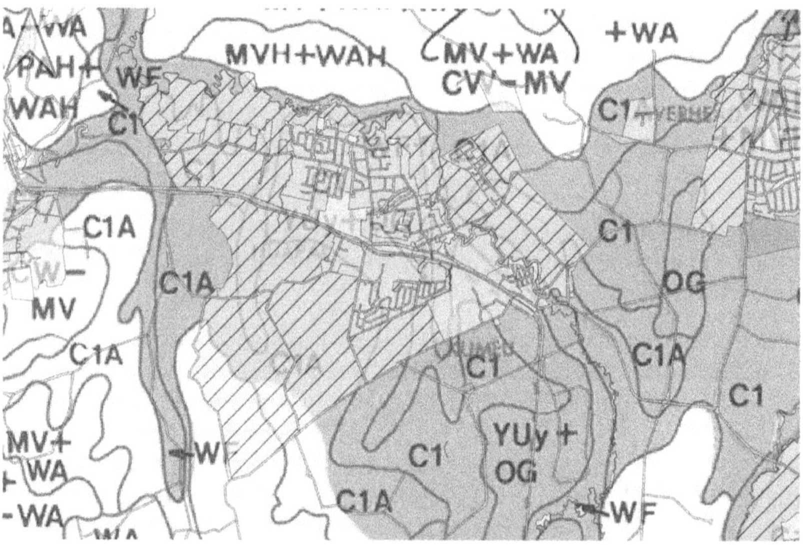


Auckland Unitary Plan for Kumeu-Huapai overlaid onto 1982 ARA High Value Soils map

*Figure 3.5* Auckland Unitary Plan zoning overlays for Kumeu-Huapai (Auckland Council, 2021) placed over the 1975 and 1982 ARA maps. The hatched section is the "Future Urban Zone", rural land awaiting rezoning to allow urban development in the future.

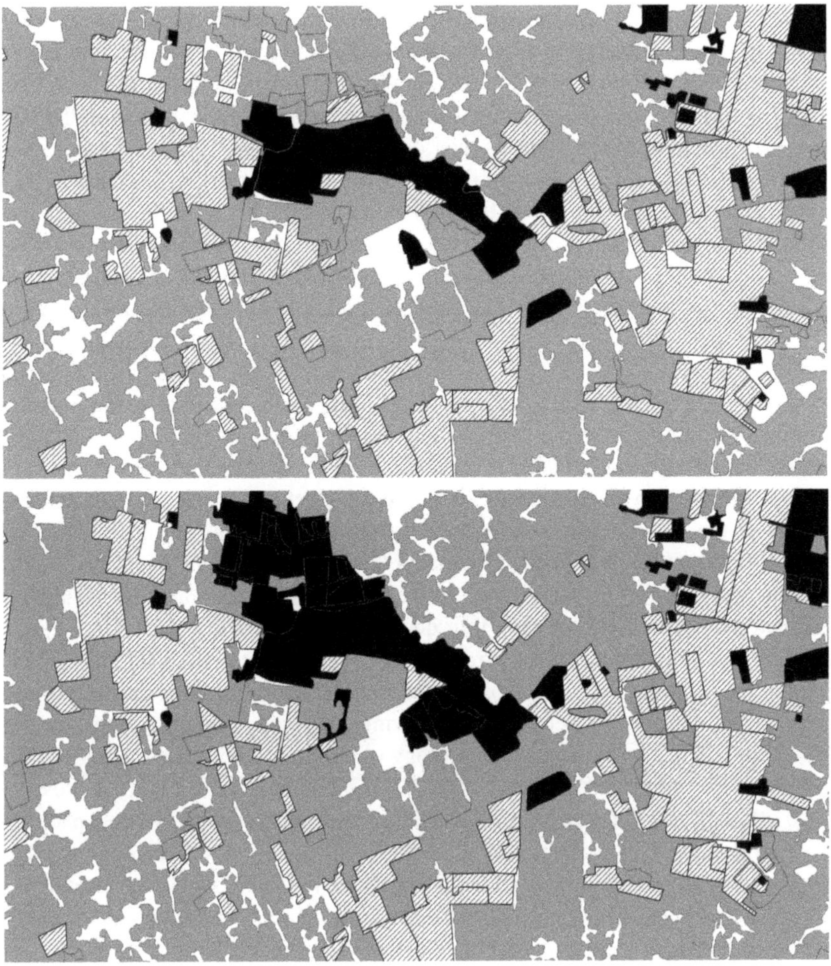

*Figure 3.6* The author's interpretation of land cover records for Kumeu-Huapai in 1990 and 2016 (Ministry for the Environment Data Management Team, 2020). The black area is urban development, and the cross-hatched areas are those that appear to be productive operations, with grey representing all other grassland areas. Note that most of the orchards are still listed as existing in the second panel despite the generalized collapse of commercial horticulture by 2016.

time. Land cover records from 1990 onwards exist and are available online (see Figure 3.6), but these simply show the apparent form of use for land as seen from aerial photography over the last few decades, rather than more fine-grained assessments of the land. The real estate agent's prediction in the first chapter that I could "drive up some driveways and find out, Huapai to Foster Road, no one's picking apples around here ... now the families are fighting over what to do with them" is a particularly dramatic expression of how the simple appearance of a landscape belies its actual formation and the social relations involved in its own "structured permanence". To discuss how the axiology of peri-urban land has changed over time, I must begin with the arguments made in previous chapters and with the systems of value employed in debates over development today.

One of the very few clues about the sharp difference in the classification of land uses in Kumeu-Huapai between 1975 and today came in the form of some remarks about the pace of suburbanization from an older resident who had long been active in demanding community participation in the development process. When I brought up the phenomenon of the rapid pace of sprawl leading to an occlusion of the productive nature of peri-urban land, as growers sell up and land bankers wait for the right moment to sell to a developer, they told me the opposite had also occurred in that brief interregnum of centralized production that made Kumeu-Huapai a horticultural centre. "Under the old Rodney Council, planners actually overestimated the value of soils in the area, perhaps to keep the place rural" they told me, going on to place this past ideological project over the landscape – the periphery as a productive resource for the city – as a mirror to the current one favouring privatized urbanization of rural areas. A Council planner and an environmental scientist working for the central government each agreed that valuations had been conducted differently in that time four decades past, but little else remains as evidence of this intentional strategy from local government to make peri-urban land literally more productively valuable by shaping the definitions that guided policy.

What this rumoured bias towards preserving rural land does offer to my analysis is a useful contrast to the dominant axiology of productive land that determines development decisions in Auckland today. Having explained how land uses are classified officially, I will detail in this section how the metrics around what constitutes productive land are flexible and subject to the position of state and landowner interests. As local government is reputed to have once overstated the relative productive value of peri-urban land to protect small-scale horticulture or perhaps even simply a rural aesthetic, systems of governance and development interests today strive (successfully) to present an image of the urban periphery as a barren landscape in need of urban development.

Speaking to one of the directors of Cabra over the phone in the middle of one of Auckland's recurrent COVID-19 lockdowns, my questions about the company's view on urban sprawl were unsurprisingly met with assurances about best practices being followed. But I was then told "the land at Huapai isn't very productive, none of the developments we did was on anything significant" followed by a statement about the broader Future Urban Zone around Kumeu-Huapai (where, as I said in Chapter 1, a local government official said he received a phone call from this same director asking about the provenance of the land). "There's not much productive agriculture going on out there now". The latter statement is entirely true today, as many real estate agents, local officials, and former horticulturalists were eager to announce whenever I asked if they knew anything about the state of the horticultural industry in the area. The former claim, however, is not only false (and genuinely upsetting to some interlocutors when I brought up the director's statement) but an attempt to assert a new understanding of the landscape's past and future – one being built by Cabra-contracted labour and machinery – over the old one.

A similar discursive move towards presenting rural land as not only currently unproductive but as only ever being suitable for urbanization is notable in legal battles over development across peri-urban Auckland. One important part of the role played by the many experts who must be called upon by landowners and developers in order to present their case for rezoning land is that of presenting value judgements on the relative worth of said land should be it developed compared to it being left as is, and it is here that definitions of productivity come under scrutiny. In a significant court case, which I have been told will somehow be relitigated soon despite having already been brought before New Zealand's High Court, landowners sought and failed to rezone the Pukaki peninsula for suburban development (see the small areas of Class 1 land on the western coast of Auckland north of Pukekohe in Figure 3.2). Even as their case was rejected, the applicants won several legal victories that have been picked up by other landowners and developers across Auckland. The High Court ruled that the Environment Court's provision about protecting the "Elite and Prime soils" discussed earlier could only ever be contingent on the land's significance for food production (The Environment Court at Auckland, 2020, p. 3), that the potential productivity of land could never be considered, only the current state of commercial horticulture on the land at the time of the application (p. 11), and that no decision on land uses should be allowed to consider the cumulative effect of the loss of many such areas of land to development over time, demanding evidence only be concerned with the exact area under discussion for rezoning (p. 20). These three points, now cited by applications for development at the furthest frontier of suburban Auckland (Envivo, 2020, p. 5), prevent the hearing of argument about what the productive peri-urban landscape once was, what it could be, and how development in one area reflects a broader transformation of this landscape. As one lawyer I interviewed put it, "the best precedents I have are from cases I lost!".

In such a framework where the only valid protection of productive land from the far more lucrative use of suburban residential development is its current commercial viability, the collapse of peri-urban horticulture and its replacement by the business of property development detailed in Chapter 1 are self-perpetuating, so long as the state continues to enable such a position on land use values. Formerly, rural towns across the periphery of the growing city, subject to similar social and economic forces as seen in Kumeu-Huapai, saw the same slow collapse of small-scale production and a move towards land banking and fragmentation. Experts working on behalf of prospective developers can then say, as put in a recent application for Patumahoe (just south of Pukekohe). "The subject land is fragmented and constrained its current use, and potential in future, making the Rural Production zoning inappropriate and out of date. A more sustainable, zoning for urban purposes is now sought" (Envivo, 2020, p. 1) and that "soils on those properties are not utilised in a productive manner beyond general stock grazing and raising domestic poultry" (p. 6). It is not an exaggeration, therefore, to say that the conversion of formerly productive land into new suburban and exurban developments acts exactly as a domino effect or, in the imagery given by one of the remaining horticulturalists, as a well-executed military manoeuvre – "some battalion lets the enemy through and surrounds us". Land that is near a successful application for development or in a Future Urban Zone and thus will inevitably be considered for rezoning in the coming decades (Auckland Council, 2019) can be made available for urbanization by pointing to the precedent being set across the periphery and thus further enable the same speculation and fragmentation that provided justification for the development.

"There's a potential for a cascade effect" one local official told me. Using the example of Cabra and Fletcher both running development projects in the area, they predicted that if one of them managed to successfully bypass regional laws and receive approval from central government to begin building suburban lots – and thus take up some of the limited development capacity still available in the Kumeu River Valley as detailed in Chapter 2 – we will see a rapid accession to the plans of other landowners and developers in the area as they fight for the remaining land they can profitably suburbanize. Through this exertion of a combination of juridical and financial power over the possible uses for peri-urban land, places like Kumeu-Huapai are transformed from places that possess a unique combination of advantages to enable food production for urban populations to suburbs in waiting, speculative investments of empty paddocks, and splintered glasshouses awaiting a massively profitable rebirth as an urban landscape with a rural aesthetic.

## Valuing land and housing at Auckland's urban periphery

The system of valuations that determine the rapidly appreciating and speculative price of land in Kumeu-Huapai is just as relational and complex as an axiology that focusses on the river valley's high suitability for horticulture, even if the former is far more occupied with that single aspect of maximal profitability at the end of any analysis. That the unprecedented rise of the price of a house in Kumeu-Huapai by nearly a million New Zealand dollars in five years occurred in the same period as the massive acceleration in development and population growth, not beforehand, is a very clear indication of how market interests react to the expansion of the urban periphery into rural areas. As much as profit-making from peri-urban development is a "speculative" venture, the appearance of certainty in the future of said development, or indeed the appearance that it already exists, is crucial. The actual sale of land from the speculatively inclined landowner to a developer intending to carve out a new suburb is not a major part of the cost of operating, even if receiving such a lump sum is doubtless transformative for the land banker, "the actual transaction costs of the land are fucking marginal … I get Chinese developers who don't even blink at the cost" one development expert exclaimed. They would later give an estimate that buying the rural land takes up approximately 2 per cent of the project's budget if one includes not just land development but building the actual houses and attendant infrastructure. As discussed in the first half of Chapter 1, developers in Auckland, with the exception of Fletcher ("because we're cradle to grave", the CEO of Fletcher Residential says to me), simply buy this rural land, reshape the earth, and lay down the supporting skeleton for houses to be built, and then sell these individually through housebuilders who seek a customer willing to buy the land and pay for the house to be built all as one package. In this process then, there are two separate financialized axiologies of land in Kumeu-Huapai – that of formerly rural land at the urban periphery and of a brand new suburban house on a ~800 $m^2$ section. These two value systems may revolve around the same end product, but have different concerns for a profitable transaction.

A property valuation expert who has worked in Kumeu-Huapai since 2013, just before the rapid expansion from 2014 onwards – "I remember when I was valuing shops on the main road and it was just a small place … there were just no new developments" – explained how they valued "empty" land in the face of encroaching urbanization. If rural land is officially zoned as being only suitable for rural purposes yet there exists a notable chance that this could change given the right pressure and connections, then "you can't just pretend it's that zone". Valuers instead adopt a "Chance of Change" methodology, where the insured or offered value of peri-urban land has the possible increase in value should it be rezoned divided by the probability of the rezoning occurring already built into it. Such a probability is, of course, drastically affected by the successful rezoning of rural land nearby, along with the general attitude of the state to rezoning, and this corresponding rise in speculative values is an indication to any enterprising financier to buy while the values continue to trend upwards.

Should the maximally profitable mode of suburban development prove out of reach, a safe profitable fallback – being considered by Fletcher and some other land developers should their bids to enable suburbanization fail – is to split the land into 1 ha lifestyle blocks. This is legally allowed if the owners purchase Transferable Title Rights from agents who have converted land elsewhere in the region into native bush (The Land Vault, 2016), allowing the protection of less valuable land further from the City to be exchanged for fragmentation at the urban periphery. Land in formerly rural areas such as Kumeu-Huapai is therefore valued (financially) in relation to a series of interests and structures that each seek to profitably employ labour and materials to sell rural Auckland to the city, and the valuations made on this basis serve to inscribe the financialized suburban landscape over the old one, long before the first surveyors and labourers arrive.

The outwardly negative aspects of the Kumeu River Valley's geography – not least the entire area's proclivity for regular flooding – are also part of this equation, but not in the way I had expected when I began this research. It is, as they tend to say in some finance circles, "Priced In". Risk of flooding damage, the valuation agent told me, is "already implicit in the value of the property", and this provided a useful competitive advantage for the area. Development in Albany has reached such a level of residential urbanization that, according to the valuer, "there's no land left in the north shore, especially for commercial-industrial, so they move here". So long as insurance continues to pay out for those who operate businesses in the area as they did after the sudden and dramatic flooding that occurred during my fieldwork, no real damage is done to the speculative possibilities of the land. "I valued some [*properties*] that flooded after they flooded … insurance just redid it, made the places better". The attitude of insurance companies towards the flood-damaged residential properties was somewhat more hostile (a relationship that will be examined in more detail in Chapter 8), but all the interlocutors I spoke to who had been affected by the flood did eventually receive complete cover from their insurer. This somewhat tense guarantee that the damage wrought by once-in-30-year floods will be repaired without cost to a landowner or tenant undergirds an abstract but significant part of the valuations for the river valley's land. Kumeu Centre, the town's only dense urban-style development, sits directly at the banks of the river, while one of Fletcher's future development sites encompasses, quite literally according to Derry "both banks of the river" (see Figure 3.7). Should the definitions of what is considered safely insurable land in the Kumeu River Valley shift, the ability to sell it profitably for at least residential land will suffer. Speaking over the phone to a Council planner familiar with the area after the flood, I was told that the anti-flooding infrastructure built for the Kumeu Central site had "worked, just. It was at the level [of the new reinforced bank], but it held".

The valuation factors that define the rapid rise in the price of a house and section in Kumeu-Huapai are as complex and distinct again from that of undeveloped land, as I would see for myself. A few days after I attend an open home for a newly built house in the Tapu/Matua Road development (see Figure 3.8), I received a phone call from the real estate agent, who must have been calling every attendee who had left their details. He asked how much I think the place is worth after a cursory request for feedback, so I thought back on previous sales in the area and said 1.45M. He replied that this price would be acceptable for a normal home, while this place – 4 + 2 bedrooms, as in a four-bedroom house with a detached two-bedroom flat – would most likely sell for over 1.6M, something that "most other responders have said" in response to his question. When I asked how all of these other respondents reached this number, he points to a neighbouring house, also classified as a "home+income" for its separate living space that could be rented out, sold for 1.4M that March. First, that house sat on a section of only 600 m$^2$ compared to the 790 m$^2$ on which the house that was the subject of this phone call sat on. Second, it had sold in March, and it was August now. Between March and August 2021, the median house price in the Auckland region had increased from an already unprecedented $1.12M NZD to $1.2M NZD (REINZ, 2021a, 2021b). A significant rise in the asking price can only be expected.

The auction for this house was held over Zoom videoconferencing due to the COVID lockdown in Auckland at the time. The auctioneer did his best to carry over the rapid pattern of encouragements that formed the core of his performance when able to conduct such a significant sales process in person:

> this is one of those locations, so close to here, that are just going from strength to strength … property can be withdrawn from the market, it happened a few times yesterday … what I love about the property, brand spanking new … you've also got the extra income … this is fast becoming a location that is so popular in west Auckland.

His speech, as skilful as it was, failed to receive an initial bid as the dozen or so attendants (most of them apparently agents for parties who deigned not to attend

*Figure 3.7* The banks of the Kumeu River next to the Kumeu Central development project (above) and winding through the Taupaki dairy farm that Fletcher purchased during the course of my fieldwork (below). Photos taken by the author.

*Figure 3.8* The newly built home, lawns still bare of any grass, with the dining room set to suggest a dinner party. Several other similar houses are visible over the fence. Photo taken by the author.

themselves) wait for one another to make the first move. To begin the process, then, the auctioneer starts the bid at 1.2M, and participants indicate a raise (beginning at 250K and then smaller intervals as the competition reaches its latter stages) with the same slight hand gesture as they would use in person. When bidding briefly stalls, the auctioneer reminds the bidders "all I wanna see is another twenty-five [thousand] for your home and income". Bidding ends at 1.615 million New Zealand dollars, but as we sit and wait for the process to be wrapped up, one of the participants has sent the auctioneer a private offer, setting off a second round of finely differentiated bidding (reaching differences of thousands of dollars rather than hundreds of thousands), before this dark horse candidate makes the initial winner relent with an offer of 1.655M. The Zoom call ends with the auctioneer thanking us all for attending and indicating to the winner that they will speak over the phone about the details. Many other houses in the area, I am told, have been sold by dint of a private offer, but as a potential customer I am assured that "there's no negotiation for a pre-auction offer, it has to be their best price", meaning the private bid would have to be significantly higher than anything the sellers could expect to see at any auction.

A few particularly relevant aspects of the axiologies employed towards peri-urban housing stand out from this entire process. This house, its detached flat, and the section of land it sits on were sold in advertisements (see Figure 3.9) and the

*Figure 3.9* A "rental appraisal" supplied by the real estate agency in the advertising material for the open home, providing a predicted rental income for any prospective landlord seeking to buy the property. Street address edited out for the purposes of this book.

auction itself on the following points of value: the separate living space (described as an "income"), a location that is "just going from strength to strength" (in the words of the auctioneer), and that it has no previous occupants. All three of these selling points are explicitly financial, targeted towards buyers evaluating the subject of the auction as an investment; as discussed in Chapter 1, investors are particularly attracted to newly built houses due to an almost total lack of maintenance costs for the first few years, while the rapid and sustained urbanization of the area promises an equally fast appreciation of the section's value. Not only is the second separate flat specifically referred to as an income, but the advertising material shown in Figure 3.9 explicitly promises that renting out the two-bedroom unit will help a buyer cover the costs of the mortgage.

Another useful point from the first chapter's analysis of the growth of property development in the area is that these financialized value judgements are not necessarily held by anonymous holding companies or even individuals specifically setting out to become property investors. In the same way that one homeowner in Chapter 1 had said they were moving out of their Huapai Triangle residence after two years in order to trade in on the property's rapidly appreciating value to find a larger house nearby, participants in Auckland's peri-urban housing market are driven towards what Gudeman (2016, p. 93) would describe as an abstracted and calculative approach to their new houses position in their lives, prioritizing financial concerns over less lucrative aspects of household and community relations in Kumeu-Huapai in the face of such undeniable profitability. With the aid of banks, homeowners in Auckland have even become ever more engaged in a "meta-financial" (Gudeman, 2016, p. 19) approach to housing by leveraging properties as collateral and relying on both continuous access to cheap credit and the unrelenting rise in house prices to allow continued acquisition of more real estate assets (Rehm & Yang, 2020). Even before the rapid increase in house prices after 2020 as investors fled to housing and other assets as investments of last resort in uncertain financial markets, this particular class of homeowners accounted for 40 per cent of New Zealand's household debt, despite making up only 8 per cent of households (Rehm & Yang, 2020, p. 91). Under such an abstracted system of valuation, the house that I had viewed and then attended an auction for – which had in fact originally been built for an investment company along with three other properties in their portfolio in Kumeu-Huapai – could not avoid being weighed and measured as a financial vehicle of potentially unmatched profitability before any other consideration, a phenomenon that applies equally to any of the other new houses that are built on the land of the Huapai Triangle and other sites of the new suburban frontier.

## FARMLUC and the NPS-HPL: presenting alternative axiologies of land in response to the financialization of urban development

As much as I have sought to emphasize how abstracted and calculative the ruling attitudes to peri-urban real estate are, to claim that the land of the urban periphery of Auckland has become entirely valued through a financialized axiology would only contribute to the same work of "making" the new suburban landscape over the old productive one that development interests seek to enact. In response to the very visible fragmentation and then suburbanization of productive rural land in Auckland, policy-makers and environmental scientists over the past decade have taken the existing structures of quantitative land valuation systems such as Land Use Classification and made their cases as to how they could be interpreted and applied better. Of particular note in this section and indeed throughout much of this project is the ongoing work of Dr Fiona Curran-Cournane, who has studied the issue of suburban sprawl and the collapse of rural productivity in the Auckland region through interviews with farming communities (2016), historical analyses of the urbanization of market gardens (2013), and broad discussions of national policy on land use (2018). Curran-Cournane's conclusions have been filed as evidence in high court proceedings over the future of development (The Environment Court at Auckland, 2020, p. 13) as one of the main counter-arguments to dominant axiologies that envision the peri-urban landscape as a barren land bank waiting to be transformed into a new suburb of Auckland City.

The initially very simple responses presented by Curran-Cournane and their co-authors to the financialized values that have been presented in the previous section – that the cumulative effect of fragmentation and urbanization be taken seriously, that growing land close to the city is valuable for its potential productivity, and restrictions on development based on historical labour and biophysical properties of the land are feasible – would radically redefine what land at the urban fringe is actually for and have been rejected in favour of the legal precedents established earlier in this chapter. What has been created from such research efforts, however, are a series of recommendations and policy proposals that, just as the 1975 and 1982 ARA maps offered a glimpse of an alternative axiology of land in the past, present a way to value productive land today and in the future that operates across and against dominant systems of valuation. Two results of this approach will be examined in this final section of the chapter: the concept of a farm-scale Land Use Classification system (abbreviated to FARMLUC) and a recently published National Policy Statement for the protection of Highly Productive Land (NPS-HPL). As a technical report and a piece of new and much-debated legislation, respectively, both of these documents have yet to have been expressed materially, but they do allow us to imagine a new landscape for the formerly rural and formerly productive towns at the periphery of Auckland's expansion, a new "Structured Permanence" defined by the often-occluded labour that made the periphery productively valuable in the first place.

FARMLUC accomplishes what one would expect from an analytical model with such a name: it reinterprets widely accepted LUC categories at the scale of a single property or productive area. The mapping of soil types past and present depicted in Figures 3.2–3.5 is conducted at scales of 1:50,000, thus removing a large amount of detail about very productive land found in some parts of individual properties or settlements (Hicks & Curran-Cournane, 2017, p. 5) and with definitions of soil conditions and limitations that suited such a heavily zoomed-out understanding of Auckland's land. Hicks and Curran-Cournane proposed to remedy these issues and thus drastically realign quantitative representations of peri-urban land by redefining subclasses and limitations for consistency across the region (remember that two identical areas of land north and south of Auckland will currently be classified differently under the standing survey data) and to apply them individually to an area of land that is being measured on its relative productive value in the face of fragmentation and urbanization. When farm-scale mapping is conducted, many areas of land that are marked as Class 2 and thus easily subdivided once local horticulture became uncompetitive are instead classified as being mostly Elite/Class 1 with some adjacent less-arable sections (p. 8), a change that goes the furthest in explaining why Kumeu-Huapai, a palimpsest of land transformations and soil types, may have been generously considered Class 1 in 1975 but not so under later regional appraisals.

Working at a finer level of detail with Land Use Classification also allows more clarity over limiting factors to productivity, allowing FARMLUC to add additional lettered subclasses for the original limiting factors shown in Table 3.2. For example, soil quality now has a more detailed list of causes, such as nutrient deficiency, toxicity, structure, and salinity, while erosion has subcategories for landslides, gullies, or slumps (p. 12). The effective results of such a substantive and small-scale approach to categorizing land in testing carried out between 2011 and 2015 are a more detailed – and generally more positive – understanding of the productive potential of land around Auckland, and a clearer picture of how the loss of individual farms and orchards to suburbanization leads to costs greater than the conclusions drawn from a 1:50,000 map or an aerial land cover survey would seem to suggest. By making a point of examining the relationships between biological, geographical, meteorological, and human material processes that make land valuable in ways beyond the speculatively financial, a system like FARMLUC can help us better value our land, in both meanings of the term.

Such an axiology, even if it provides a clearer understanding of the productive value of rapidly fragmenting land at the urban periphery, would need a corresponding set of commitments from governing forces, through an alignment of ruling interests, in favour of protecting small productive operations at the edges of the city. As has been established in this and every previous chapter, this has not been the case for nearly half a century. The High Court has established that the cumulative loss of productive land or the potential productivity of a collapsed orchard cannot be considered in land use debates, and state support for horticulture at anything below an industrial scale has waned as the business of exurban and suburban property development has risen. A more sustainable axiology of land would have to come from the very top, a demand made by Pukekohe farmers now (Corlett, 2018) and bitterly remarked upon by former horticulturalists in Kumeu-Huapai. To this end, and in the face of concerted media coverage of peri-urban sprawl onto productive land in Auckland (Dreaver, 2019; Radio New Zealand, 2019), a draft for a National Policy Statement to Protect Highly Productive Land was drawn up in late 2019 (Ministry for Primary Industries, 2019). Central government would create and enforce definitions of productive land that would be protected from fragmentation and development in order to enable the existence of peri-urban horticulture in the Auckland region for the foreseeable future, or at the very least mitigate its collapse in the few remaining and crucially important productive areas. Criticism from the Treasury that this would restrict urban growth and thus slow one of New Zealand's greatest economic engines (p. 19) was allayed with promises of using this for "redirecting rather than constraining growth" (p. 2).

This draft became a published policy statement three years later, in September 2022. In its current form, the statement requires each regional authority to identify land under their jurisdiction that is highly productive land, protect it from further subdivision, and involve Tangata Whenua in doing so (Ministry for the Environment, 2022). Crucially, all of the productive land in Auckland's Future Urban Zones is not subject to this process, and the consequences of ceding this land to development will be elaborated on in the following chapter. In the intervening years between the presentation of the draft policy and its publication, other National Policy Statements that place pressure on local authorities to release more land for development have been written, discussed, and passed by the majority

Labour government (New Zealand Government, 2020), a move intended to, as one official told me "flood the market with capacity". At present, any much-needed realignment of the state towards a less financial evaluation of land in places like Kumeu-Huapai is not forthcoming.

The formerly productive land of the Kumeu River Valley is today categorized uniformly as Class 2 land, and with the collapse of local horticulture and the meteoric rise of suburban development, such a classification has had little effect on the shape of the area's rapid urbanization. Some faint traces remain of an older approach to land use which placed the area in greater esteem, and attempts to create a more comprehensive axiology of the potentials of peri-urban land are being created in response to the increasingly obvious financialization of land relations, but neither of these axiologies (or indeed ontologies) are aligned with the way that the group of people who we often label as state and market forces approach urban development in Auckland today. In order to understand how this particular collection of individuals, institutions, and interests does understand the land (beyond simply as a suburb-in-waiting, as I have described in this and previous chapters), how they enable and benefit from the unending demand for land at Auckland's urban periphery, and who suffers from this from this state of affairs now and in the future, the following chapter examines the political economy of Auckland's urban development. I will detail exactly how landowners, developers, local officials, and central government each relate to the industry of residential development, that single sector of industry that underpins New Zealand's entire financial urban economy.

## Bibliography

Auckland Council. (2019). *Decision Following the Hearing of a Plan Modification (PPC25 – Warkworth North) to the Auckland Unitary Plan under the Resource Management Act 1991*. Retrieved from https://www.aucklandcouncil.govt.nz/UnitaryPlanDocuments/pc-25-decision.pdf

Auckland Council. (2021). *Auckland Unitary Plan Operative in Part*. Retrieved September 10 from https://unitaryplan.aucklandcouncil.govt.nz/pages/plan/Book.aspx?exhibit=AucklandUnitaryPlan_Print

Auckland Council. (2022). *Auckland Unitary Plan, Operative in Part – Chapter J: Definitions*. Retrieved May 24 from https://unitaryplan.aucklandcouncil.govt.nz/Images/Auckland%20Unitary%20Plan%20Operative/Chapter%20J%20Definitions/Chapter%20J%20-%20Definitions.pdf

Corlett, E. (2018). Pukekohe grower: 'We're really the last frontier'. *Radio New Zealand*. Retrieved August 15 from https://www.rnz.co.nz/news/national/364134/pukekohe-grower-we-re-really-the-last-frontier

Curran-Cournane, F., Cain, T., Greenhalgh, S., & Samarsinghe, O. (2016). Attitudes of a farming community towards urban growth and rural fragmentation—An Auckland case study. *Land Use Policy 58*, 241–250.

Curran-Cournane, F., Golubiewski, N., & Buckthought, L. (2018). The odds appear stacked against versatile land: Can we change them? *New Zealand Journal of Agricultural Research 61*(3), 315–326.

Curran-Cournane, F., Vaughan, M., Memon, A., & Fredrickson, C. (2013). *Auckland's elite and prime land: Similar messages and continued trade-offs 54 years later*. Auckland: Auckland Council Technical Report 2013/050.

Dreaver, C. (2019). Environment minister: 'We actually prefer potatoes'. *Radio New Zealand*. Retrieved August 14 from https://www.rnz.co.nz/news/political/396668/environment-minister-we-actually-prefer-potatoes

Dunsford, D. (2002). *Doing it themselves: The story of Kumeu, Huapai and Taupaki*. Auckland: Kumeu District History Project.

Envivo. (2020). *Private Plan Change Request to the Auckland Unitary Plan: Operative In Part Patumahoe South Plan Change Request Assessment of Effects and Section 32 Assessment*. A. Council. Retrieved June 26 from https://www.aucklandcouncil.govt.nz/UnitaryPlanDocuments/pc-55-plan-change-request.pdf

Google LLC. (2022). *Google Satellite via QGIS*. Retrieved from http://mt0.google.com/vt/lyrs=s&hl=en&x={x}&y={y}&z={z}. Source: http://mt0.google.com/vt/lyrs%3Ds%26hl%23Den%26x%23D%27Bx%27D%26y%23D%27By%27D%26z%23D%27Bz%27D&zmax=18&zmin=20. Provider: wms.

Gudeman, S. (2016). *Anthropology and economy*. Cambridge: Cambridge University Press.

Hicks, D., & Curran-Cournane, F. (2017). *Matching farm production data to land use capability for Auckland – Technical report 2017/020*. Auckland Council Research and Evaluation Unit, RIMU. Retrieved from https://www.knowledgeauckland.org.nz/publications/matching-farm-production-data-to-land-use-capability-for-auckland

LRIS Informatics Team. (2011). *LRIS Data Dictionary Version 3*. Retrieved September 27 from https://lris.scinfo.org.nz/document/9162-lris-data-dictionary-v3/

LRIS Informatics Team. (2021). *NZLRI Land Use Capability 2021*. Retrieved from https://lris.scinfo.org.nz/layer/48076-nzlri-land-use-capability-2021/

Ministry for Primary Industries. (2019). *Proposed National Policy Statement for Highly Productive Land – Cabinet Paper*. Retrieved from https://www.mpi.govt.nz/dmsdocument/37065-Proposed-National-Policy-Statement-for-Highly-Productive-Land-Cabinet-paper

Ministry for the Environment. (2022). *National Policy Statement for Highly Productive Land – September 20 2022 ME1682*. Retrieved November 11 from https://environment.govt.nz/publications/national-policy-statement-for-highly-productive-land/

Ministry for the Environment Data Management Team. (2020). *LUCAS NZ Land Use Map 1990 2008 2012 2016 v008*. Retrieved May 26 from https://data.mfe.govt.nz/layer/52375-lucas-nz-land-use-map-1990-2008-2012-2016-v008/history/

Mitchell, D. (2003). Dead labor and the political economy of landscape – California living, California dying. In M. Domosh, S. Pile, & N. Thrift (Eds.), *Handbook of cultural geography* (pp. 233–248). Thousand Oaks, CA: Sage Publications.

New Zealand Government. (2020). *National Policy Statement on Urban Development 2020*. Retrieved from https://www.hud.govt.nz/urban-development/national-policy-statement-on-urban-development-nps-ud/

Radio New Zealand. (2019). *Building on Prime Growing Land – Nine to Noon*. Retrieved May 2 from https://www.rnz.co.nz/audio/player?audio_id=2018693232

Rehm, M., & Yang, Y. (2020). Betting on capital gains: Housing speculation in Auckland, New Zealand. *International Journal of Housing Markets and Analysis*, *14*(1), 72–96.

REINZ. (2021a). *August 2021 Property Reports*. Retrieved from https://www.blog.reinz.co.nz/reports-1/august-2021-statistics

REINZ. (2021b). *March 2021 Property Reports*. Retrieved from https://www.blog.reinz.co.nz/reports-1/march-property-reports

The Environment Court at Auckland. (2020). *Decision No. [2020] NZEnvC 214 in the matter of the Resource Management Act 1991 and the Local Government (Auckland Transitional Provisions) Act 2010 and of an Appeal under Section 156(1) of the Local Government (Auckland Transitional Provisions) Act 2010 between Self Family Trust (ENV-2016-AKL-000199), Apellant, and J & F Gook, Section 274 Party, and Auckland Council, Respondent, and Ted Edwards, Section 274 Party.* Retrieved from https://environmentcourt.govt.nz/assets/Documents/Publications/2020-NZEnvC-214-Self-Family-Trust-v-Auckland-Council.pdf

The Land Vault. (2016). *TTR's Explained.* Retrieved June 2 from https://www.thelandvault.co.nz/transferable-ttrs-explained/

# 4 The political economy of peri-urban sprawl in Auckland

*Figure 4.1* Real estate services advertisement sign on Kumeu-Huapai's main road, in front of the rail line. Photo taken by the author.

In interviews where I discussed the legal debates over peri-urban sprawl, representatives of regional authorities and regulatory bodies tended to have a particularly antagonistic view of the development industry and its attendant trades. "Developers take pride in running circles around council, there are planners that specialize in fucking up the unitary plan … we've created an industry of planners who make a living in arguing the toss", one local official told me despairingly. The ill feeling is mutual, though developers most often depicted governance bodies as incompetent and sclerotic rather than intentionally obstructive, the managing director of a land

DOI: 10.4324/9781003400325-6

development company saying "Council and the Ministry of Environment need to plan better than they do, they have plenty of time to plan and they've done nothing" in response to my questions about the issues caused by the rapid suburbanization of the area.

As formerly rural towns such as Kumeu-Huapai double in population in a matter of years and real estate becomes ever more the final refuge of safely profitable investment, the endless demand for land at Auckland's urban periphery to be released for residential development has become a central site of conflict in the City's political economy. Even a very broad picture of the totality of New Zealand's economic output shows that property is the biggest moneymaker in the country (see Figure 4.2), and it is in the cities of Aotearoa, particularly Auckland, where what Jane Kelsey (2015) labelled the FIRE economy (Finance, Insurance, and Real Estate) is taking an increasingly dominant position over every aspect of governance, work, and life. As the business of suburban development on rural land has become crucial to the financial success of Auckland City, policies created by local and central government in an effort to provide affordable housing while protecting ecosystem services have had to reckon with the inescapable gravity of privatized residential development over decision-making processes. As that local official had complained to me, entire industries of professionals have become notable for utilizing their familiarity with land use policy to further the ends of their clients. Indeed, as the opposing developer had claimed, planners and policymakers at local, regional, and national levels are finding themselves unable to stand in the way of the endless demands of the development sector, let alone offer an alternative vision of the future of Auckland.

This chapter is the culmination of the half of this book that has focussed almost entirely on the urbanization of Kumeu-Huapai and other towns at the periphery of

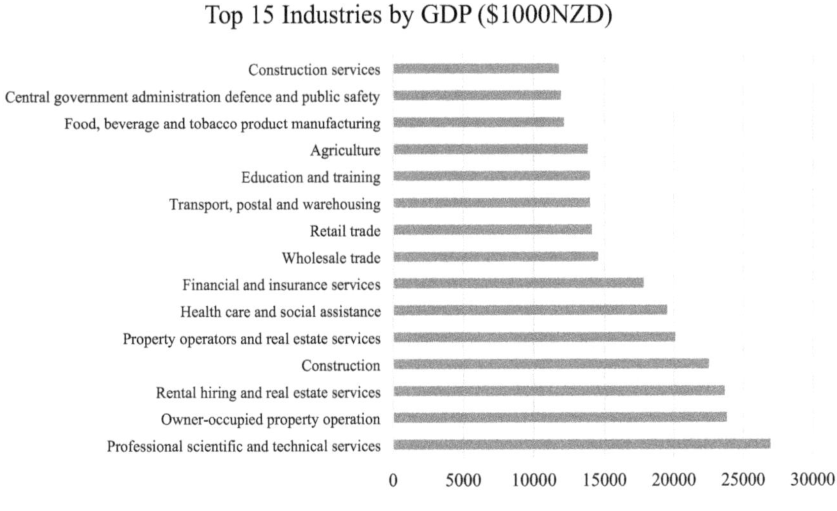


*Figure 4.2* New Zealand's top 15 GDP-contributing industries in 2020 (Stats NZ, 2020b).

Auckland's accelerating expansion. It looks at the past decade of attempts by New Zealand's central government and Auckland Council to create policies with the aim of providing affordable housing for its people while mitigating the degradation of productive land and ecosystem services and examines how such outcomes failed to be realized. I begin with an analysis of the political economic landscape in which the suburbanization of rural Auckland is occurring, showing how the relationships between landowners, developers and financiers detailed in earlier chapters facilitate urban sprawl across the region in a time of unprecedented housing insecurity. Next, I will discuss the formation and effects of a series of policies enacted in the past ten years by past and current governments in their efforts to mitigate this housing crisis. Four policy efforts will be brought up: the National government's Special Housing Areas (first mentioned back in Chapter 1), operative from 2013 to 2017, the ongoing conflict over Auckland Unitary Plan since 2016, the current Labour government's National Policy Statement on Urban Development, released in 2020, and finally the bipartisan housing reform bill (an amendment to existing resource management legislation), written and passed in late 2021. Lastly, I will detail how this confluence of interests and legislation has enabled an extremely privatized and financialized form of urban development in Auckland at the expense of all other concerns. Every level of governance that has jurisdiction over the region has found itself unable to contend with the FIRE economy and its influence on working life.

## Vested interests: conceptualizing Auckland's rapidly fragmenting urban periphery as a field of forces

As has been indicated in previous chapters, landowners – whether absentee land bankers, landlords or owner-occupiers – are not passive bystanders in the process of urban sprawl but are each active agents (Murphy, 2020) seeking to make the best of their claim over the land. Their relationships with development professionals are crucial to realizing such a windfall. It has also been established that my interlocutors did not necessarily hold a single fixed position in the urban development process. A well-connected real estate agent makes their fortunes from their position in local government, and a canny homeowner utilizes the increase in speculative valuations of the place in which they live to quickly trade their house for a larger one in the same town. To best depict the interactions between individuals who could broadly be described separately as landowners, financiers, development professionals, and public officials, but in day-to-day life engage with a multiplicity of these labels depending on their own relationships, I will return to the work done by Pierre Bourdieu (2005) to understand the French housing market. More than simply illustrating how the interests of financial institutions, the cultural dispositions of the urban professional class, and the strategies of peri-urban developers led to the dominance of the industrially constructed "traditional" mason house across the urban periphery, Bourdieu sought to depict the internal and inter-personal processes that shaped the paths of individual legal processes at this suburban frontier. To do so, he presented every area subject to its own local authority as a Territorial Field, affected by the demands of centralizing regional and national governance yet defined by how such regulations are actually interpreted and applied by individuals on the ground. He argues that:

> A field, as a flexibly structured and minimally formalized area of free play, or even a bureaucratic organization as an artificially structured game, constructed with explicit ends in mind, is not an apparatus obeying the quasi-mechanical logic of a discipline, capable of converting all action into mere execution.
>
> (p. 130)

and instead that the organizational structures that reproduce patterns of urban development do so at the wishes of the local individuals with the knowledge and capital to ensure it occurs on their terms. Kumeu-Huapai's rapid suburbanization is not merely an expression of global trends but a result of particular relationships between individuals at the edge of the growing city.

In describing the freedoms available to these privileged subjects at the urban periphery, he explains the boundaries of their agency thus:

> These "agents of execution, who are never mere executants, always have available to them a range of possible "choices" lying between two limits, though these are doubtless never actually reached: at one end of the scale is the strict, complete application of the regulations, with no consideration for the details of the case in question, which, as the phrase "summum jus, summa injuria" reminds us, may be an impeccable form of the abuse of power; or, by contrast, a legitimated transgression, an official or semi-official dispensation, in the sense of an exception to the rule made within the rules, and a legally sanctioned privilege.

In Kumeu-Huapai and in other formerly rural towns at the edge of Auckland's expansion, the material realities of land fragmentation and suburban sprawl arise not directly from the wording of policies set down by regional and national authorities but from the initiative of development agents and local government officials to define and express these rulings in ways that maximally benefit them.

The other side of this lens that focuses on individual initiative as an expression of institutional power, however, is that this power is not evenly dispersed across every actor seeking to express their reality in the field. While Bourdieu does acknowledge that the "weaker", less connected players in the game of peri-urban development can utilize the conflict between major institutional interests to "find paths towards independence" (p. 137), they will tend to find themselves much more restricted on the spectrum of strict application to legitimated transgression. Thus, an internationally owned development company can purchase rural land confident in the knowledge it will be able to convince the central government to bend the rules for it, while the patriarch of a family horticulture operation complains that his attempt to sell his own productive land for development was flatly denied. The work of profitably transforming several dozen hectares of rural land in Auckland into a new suburb of a thousand houses that must be then marketed and sold would not be possible without the "massive machine" described in the first chapter, and therefore, the Territorial Field of peri-urban development in Kumeu-Huapai is dominated by a few powerful interests and inclinations around which all others must arrange themselves. Landowners, planners, expert consultants, and other individuals charged with the execution of urban development policy at Auckland's periphery seek to make their own future for the area, but they do not do it as they please. Individual agency and intentionality – even among well-established professionals – operate in the context of the small number of powerful development companies and financiers, alongside separate local, regional, and national juridical frameworks represented in turn by agents seeking their own outcomes.

## Justifying the means: the political economy of peri-urban development amidst a housing crisis

When a representative of the party seeking to make the Pukaki Peninsula available for residential development had their moment at New Zealand's High Court to justify an expansion of Auckland's Rural Urban Boundary and the transformation of a protected area into suburban housing, his argument was entered into the record thus:

> Every sensible opportunity to provide housing at whatever scale – public rental housing, affordable housing, high density housing or suburban housing, will be of value to answer this critical, social and community need. I simply do not agree that insufficient justification for additional growth exists.
>
> (The Environment Court at Auckland, 2020, p. 141)

As noted in the previous chapter, development in the Pukaki Peninsula has been denied for now. No matter how impassioned the speech made to demand it, the representative himself admitted during proceedings the area offered "limited potential contribution" to issues of housing supply and was told afterwards that the developer's planning team had failed to assure the court that they would account for the impacts of suburbanization on the area and provide a good quality of life for residents (p. 177). This failed bid, however, reflects what was perhaps the strongest moral discourse presented to me in interviews by development professionals concerned with the public opinion of accelerating suburban development in Auckland – that no argument against further urbanization of the periphery can stand up in the face of the critical need for more housing, regardless of the form.

This posture from developers is not without basis. The median price of a house in Auckland proportional to the median income is now only surpassed by the likes of Hong Kong, Vancouver, Sydney, and San Francisco, at a "severely unaffordable" 11.2 times the median income in 2021 (Urban Reform Institute and the Frontier Centre for Public Policy, 2022, p. 8). In legal conflicts over urban sprawl and in the policy written to define acceptable land uses, the unprecedented rise in the cost of housing for Auckland residents acts as a powerful discourse harnessed by individual interests seeking to rezone the land. Applications to have an area of rural land made available for development against the wishes of Auckland Council make sure to remind the Council of their duty to provide "much-needed" housing capacity for the City, alongside the more commonplace arguments about suburban housing being the most lucrative and therefore best land use (Barker and Associates Urban and Environmental, 2020, p. 3). Interlocutors in the development sector would also often portray their work to me as a public service, unfortunately being obstructed by bureaucratic red tape and institutional inertia, "the planning they do is so slow" one director complained, another saying the blame for issues caused by suburbanization lay with authorities "housing is not the problem, poor planning is the problem". What is needed at the urban periphery, according to these parties, is for the Council to be suborned to the interests of private development.

This moral argument, that any restriction on the possibilities and forms of private residential development necessarily increases costs and delivery times for new housing, is contested and litigated by those who stand to gain from one side of the process or another in Auckland and across this urbanizing planet. A particularly strong recurring set of academic discourses around this issue have placed regulatory planning at the centre of the global crisis in housing affordability since the new millennium. Glaeser and Gyourko (2002, 2003) and Glaeser et al. (2005) continue to hold a central place in theoretical discussions of property values with their enduring argument (Diagne et al., 2018; Manville et al., 2022) that the cost of a house for a prospective buyer is mostly (often beyond 90 per cent in large cities) (Glaeser & Gyourko, 2003, p. 30) composed of an artificial scarcity created by zoning regulations. O'Flaherty (2003) and Murray (2021) presented critiques of Glaeser and Gyourko's figures and methods (Murray, 2021, p. 195; O'Flaherty, 2003, p. 41), as well as, more relevant to this chapter, concern about the political economic context that led to a position that demands that all barriers for property development interests be removed being so enthusiastically adopted by think tanks and governments in recent years (Murray, 2021, p. 207). Despite such opposition, Glaeser and Gyourko's model continues to hold a powerful position in arguments over land use governance around the world.

This paradigm is no less relevant in conflicts over Auckland's own housing crisis. The rapid and sustained rise in house prices across Auckland City in the first decade of the millennium led Grimes and Laing (2009) to point to Auckland's Metropolitan Urban Limit (MUL) as the most significant driver of unaffordability, claiming that land within this boundary was ten times more expensive than land beyond it. In a similar vein, Lees (Lees, 2019, p. 256), in an argument that Martin and Norman described as "to some extent replicating work by Edward Glaeser in the US" (Martin & Norman, 2020, p. 9), claimed that 55 per cent of property's value in Auckland was created by land use regulation. As with Glaeser and Gyourko, Grimes, Laing and Lee's assessments of the causes of housing affordability do not go unopposed. Martin and Norman's (2020) report in their capacity as Chief Economist Unit for Auckland Council argued that both papers were obsolete due to a combination of dramatic changes in land use legislation in New Zealand since the publication of these papers as well as some significant issues with their methods (p. 8), offering instead their own calculations that Auckland's Rural Urban Boundary (which has replaced the MUL) imposes a price premium of approximately 5 per cent (p. 26). This criticism aside, both of these articles are regularly cited as solid examples of the relationship between planning regulations and the sale price of a house in New Zealand today (de Roiste et al., 2021, p. 595; Hong, 2020, p. 39). Through this book, however, I seek to speak against the dominant discourse that land use regulations are the overdetermining factor for the failure to provide affordable housing to the people of Aotearoa. Recent academic literature on urban development has presented the argument that actually existing land supply has very little impact on house prices in growing cities (Wetzstein, 2021, p. 9) and in fact, as will be discussed later in this chapter, that the fast-tracking of building consents only offers more speculative value to a piece of land in the path of urban expansion (Fernandez et al., 2021, p. 152). I instead intend to depict housing unaffordability, urban sprawl, and land fragmentation as symptoms of financial relationships that benefit some parties as much as they reduce the quality of life for many others.

This ideological battle of peer-reviewed articles and professional experts plays an important role in the forms of Auckland's peri-urban development, not merely at the abstract level of writing policy but in the interactions between the individuals seeking to execute the development process in Kumeu-Huapai in a way that benefits their interests. Council planners and local officials noted that their respective regulatory organizations had a "tendency to fold under pressure" for unplanned or out-of-sequence residential suburban development to be allowed at Auckland's urban periphery. The reason, apart from policy issues to be discussed later in this chapter, was that these interpreters and executors of land use policy faced an ever-mounting demand from the central government to make as much land available for development as possible amid a context of such severe unaffordability of housing for the residents of Auckland. A local official described the relationship as an "intense pressure on council, if you don't do this, we'll do it anyway, just legislate past you".

The New Zealand Government and Auckland Council do agree that the housing affordability crisis will be sold by leveraging policy to encourage private developers to provide new housing, but the arguments begin beyond that point. For a decade, representatives of the central government in Wellington have sought above all to release peri-urban land for residential development, while Council planners have attempted to prioritize urban intensification (Fernandez et al., 2021, p. 144; Murphy, 2016). Policies created by the latter must necessarily be restrictive for developers, preventing them free reign to engage in the maximally profitable activity of suburban development on Greenfield land and demanding dense forms of residential housing which, at least according to what one land developer claimed in an interview, is double the construction cost per square metre compared to a single-family suburban house. In response, developers and landowners must use whatever connections and expertise they can muster in order to bypass or remove attempts by Auckland Council to regulate urban development in the region.

The result is the "industry of planners who make a living in arguing the toss" described in the opening of this chapter. Developers and their political representatives complain publicly that regulating sprawl defines the rise in house prices in the ways described by Glaeser and Gyourko (Hong, 2020), while experts testify in hearings over zoning laws that "the risks of undersupplying development opportunities … are greater than oversupplying them" (Auckland Council, 2019, p. 19) and that allowing land to be released for development in a manner contrary to existing plans "will help meet the Council's obligation" to provide new housing (Auckland Council, 2020, p. 57). The inclinations of planners and policymakers working for Auckland Council have become one of the main barriers for the private development sector's ambitions, and therefore a target for any supporter of private development as the means to solve the housing crisis.

"At some point, planners became bogeymen" a local official remarked, after we had discussed the local government's relationship with the development sector. While these bogeymen are often opposing suburban sprawl in Auckland for reasons more to do with their goal of maximizing the fixed capital of urban areas and thus their income from property rates than from any overwhelming anti-capitalist or pro-sustainability feeling, the combative attitude between the parties described at the start of this chapter is unavoidable. Council planning efforts are seen by property developers, landowners, and a government desperate to avoid the repercussions of the financialization of the housing market as an obstacle to be removed. In a phone call with a planner who had worked against several cases of unplanned development, they mused that Auckland Council as an institution technically possessed the ability to veto a suburban residential development and demand denser forms of urbanization, but such an open move of opposition would destroy what remained of their relationship to policymakers in Wellington and most likely prompt a direct intervention.

A lawyer who has often acted on behalf of developers presented their clients' view of this conflict:

> they [*developers*] know engineers and surveyors, they know people in council, they want to get land converted at the least cost with the least delay, nothing hurts a margin more than expenses you didn't budget more, and delays and obstructions … not only do they [*the developers*] pay for consultants, they're paying for peer review … so developers see councils as a direct threat.

For the entire sector, therefore, any slowdown, check, or restriction on the development process poses a threat to the cycle of capital accumulation based on the conversion of rural land to housing by which the surveyors, engineers, housebuilders, environmental consultants, lawyers, and most importantly financiers involved continue to make their livings and contribute to Auckland's FIRE economy. As will be demonstrated in the following section, attempts to create and enforce land and housing policy on a top-down basis in a field dominated by the relationships between private land developers and their collaborators are contingent on such policies continuing to enable the unending demand for land at Auckland's urban periphery.

## Policies to enable private development in Auckland

Derry, the local historian who first featured in Chapter 2, remembers the moment when the first and largest Special Housing Area (SHA) was announced. On the land that was once Nobilo's vineyard, Nick Smith, then the housing minister, and Len Brown, Auckland's mayor at the time, held a conference. "Nick Smith waved his hand saying 'I can see it, all this, special housing area'" Derry recalls. An official who had been involved in the process of streamlining the transition of the land that was to become the Huapai Triangle from rural to urban told me that Auckland Transport, the Auckland Council-controlled transport agency for the region, "had weeks" to calculate the transport infrastructure needs of the new development. The SHA, as with each of the three other policies to be discussed in this section, exists in a social and political formation that has ceded the provision of housing to the small group of agents in finance, development, and construction often generalized as "the market" (McLeay, 2020, p. 128). SHAs were submitted to the New Zealand Government by developers working with landowners seeking to increase the value of their land, a fact that was raised in interviews by real estate agents, Council planners, Local Board members, and residents of Kumeu-Huapai. As detailed in Chapter 1, these interlocutors explained that the Huapai Triangle was originally zoned to be industrial land, until the half-dozen or so landowners co-ordinated with developers to have the area rapidly converted into a suburb in the making. As with every other SHA that would be established in New Zealand, the release of the Huapai Triangle for urban development came with no requirement that any houses be sold at a price affordable for low-income households (Murphy, 2016, p. 2542). Instead, prices would apparently be reduced entirely by increasing the supply of housing.

At the time of my fieldwork, eight years since it became an SHA, the Huapai Triangle is not finished, and it is certainly not fully occupied (see Figure 4.3). Kumeu-Huapai residents told me that they had been promised that the SHA would "unlock" funding for new road infrastructure and public transport, an idea reflected in the proposal document for the Huapai Triangle itself (Brown & Company Planning Group, 2015, p. 5), yet no such funds or infrastructure appeared. Even for people like Chris Penk, local National Party MP for Kaipara ki Mahurangi and generally supportive of the SHA, telling me "It solved the supply and therefore affordability problem", there is no avoiding the fact that the Kumeu-Huapai has doubled in population over the past seven years with no commensurate infrastructure efforts. Chris said, "nine out of ten people that come into this office are here to complain about the state of roads". Greg Sayers, the councillor for the Rodney Ward of Auckland, was much more blunt, saying the SHAs were "probably a failed experiment", and the impacts on the community were "unwanted consequences from legislation meant to provide housing affordability". A member of the Local Board was dismissive of the entire SHA project nationwide, explaining that "none of the SHAs have been finished" and that I shouldn't expect to see any different from private developers who "like to build and release houses slowly to protect their margins". This local official would later muse that if indeed land developers

*Figure 4.3* The Huapai Triangle in August 2021. Photo taken by the author.

were compelled to create and release housing at the rate they claimed to be able to accomplish, then there would genuinely be a glut of supply for large suburban houses at the urban periphery, "if you actually built the houses, the market would look very different".

As the private development process stands today, however, no such decrease in house prices, and therefore profitability, has been allowed to occur. Despite Chris Penk's best hopes, SHAs have not only failed to create any kind of affordability effect by streamlining land supply but also caused the exact opposite effect. Across the country, SHAs create "an average housing price increase of about 5 per cent" and increase the probability of more costly housing transactions in the area (Fernandez et al., 2021, p. 144). SHAs were successful in one respect, which was ensuring individual agents in the territorial field of peri-urban development such as landowners and development professionals were able to make their fortunes by rezoning rural land on the promise that it would be quickly released for residential development.

While SHAs were shelved by the new Labour-led coalition government in 2017, this underlying dynamic – landowners and developers seeking to rezone land to build suburban housing at the urban periphery for maximal profitability – has continued to define debates, plan changes, and court decisions about land use in Auckland. The Auckland Unitary Plan (Auckland Council, 2021a) was made operative in part in 2016 and is modified continuously in an effort to meet the demand for development while mitigating its negative effects. The Rural Urban Boundary and Future Urban Zones (see Figure 4.4) first shown in Chapter 1 were created by this plan. This document, its admirable qualitative aspirations for the future of Auckland City (high-quality urban living, management of sprawl, and reversal of environmental degradation), and its refusal to make clear criteria for what these aspirations should actually look like for the people who will live there reflect the ongoing conflicts between individual parties over development and land use in the Auckland region under an unprecedented housing crisis. As described in Chapter 1, landowners and developers can make submissions for plan changes, and according to one Council planner, "they can win them easily … current narrative is that any house is good" due to the discourse of urgency and moral imperative presented by the appellants in any one attempt to secure a legitimated transgression of existing planning law. These submissions promise "to enable subdivision, use and development within a greenfield area while ensuring any adverse effects on the environment can be appropriately avoided, remedied or mitigated" (Auckland Council, 2021c, p. 44) and assure authorities that "the Plan Change will provide quality, compact neighbourhoods" (Barker and Associates Urban and Environmental, 2020, p. 5). Such plan changes are all concerned with converting greenfield areas at Auckland's urban periphery into residential developments, all because, as another Local Board representative put it, "they can build fast, sell high and leave quick", yet they are able to appeal to the Unitary Plan's demands for a best-practice approach to development by simply assuring the adjudicators of the process that they will do their best to follow such aspirations.

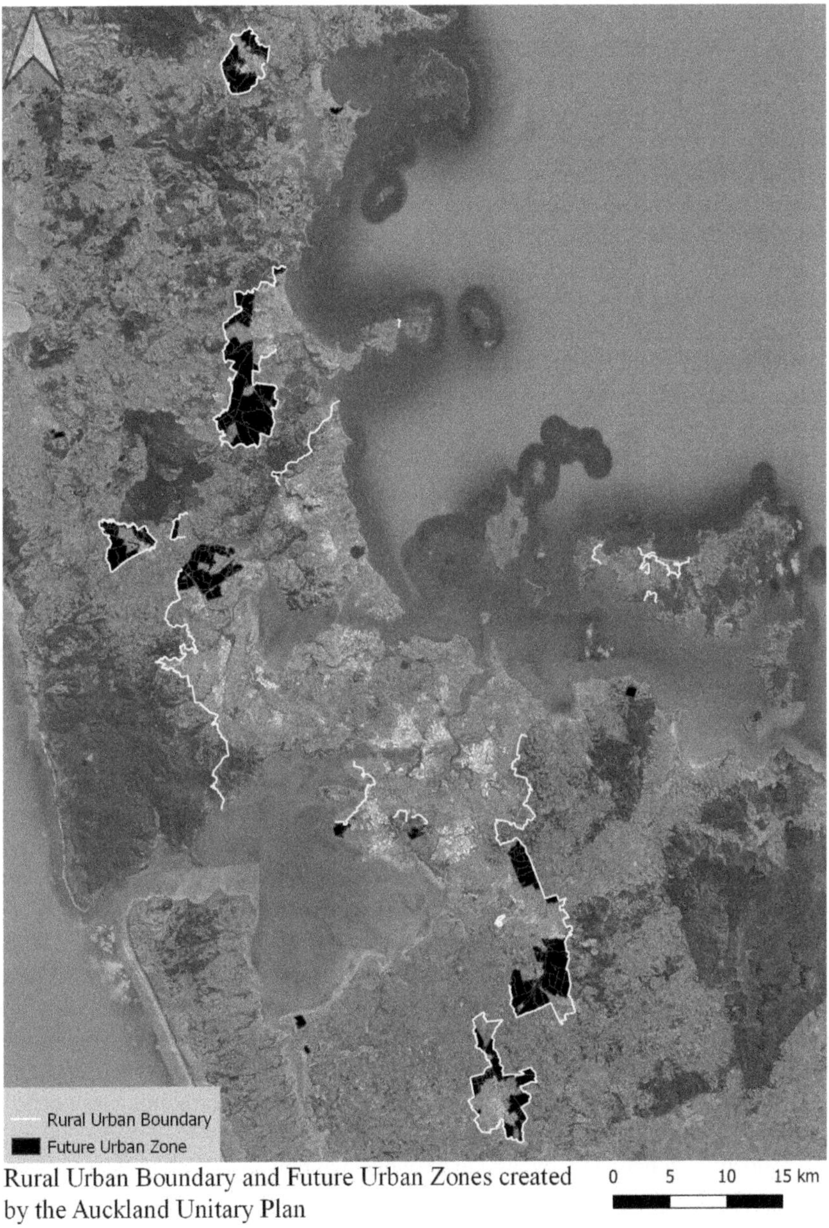


*Figure 4.4* The Rural Urban Boundary as of early 2022 marked in white and the Future Urban Zones marked in black. Zoning data from the Auckland Unitary Plan (Auckland Council, 2021a) and satellite imagery from Google Maps (Google LLC, 2022).

When decisions are made over possible changes to the Auckland Unitary Plan such as these, a panel of qualified development and planning experts is assembled, which introduces another aspect of peri-urban development defined by the relationships between the executors of policy and law and the development sector. Taking the course required to become accredited to act as a member of the hearing panel costs $2,200 NZD (WSP New Zealand Limited, 2022), and almost all involved are themselves involved in the business of peri-urban fragmentation and suburbanization. After all, where else could one find professionals with the expertise necessary to judge such cases? The result of this arrangement is that directors of the various consulting companies who work closely with land developers also act as independent decision-makers on rezoning decisions. "That's where it gets dodgy" a Council planner admitted, noting that the professionals on the hearing panels making these legislative judgements were often waiting on a subdivision or development of their own to be cleared and they would want to establish precedent, "wearing two hats at the same time". By making no attempt to go beyond the same market-driven narratives that define the political-economic context that created the SHAs, the Auckland Unitary Plan is similarly unable to improve housing and land use outcomes in Auckland.

The greater profitability of residential development compared to commercial or industrial land necessarily drives landowners and developers in Auckland's FUZ to push for changes to the Unitary Plan even if they have already been assured that their rural land will soon be rezoned. In Warkworth, the northern frontier of Auckland's urban sprawl as Kumeu-Huapai and Pukekohe are the western and southern frontiers (the northernmost centre of Future Urban Zoned land in Figure 4.4), interested parties put forward a plan change to change an area of land from "Business Light Industry" (BLIZ) to "Business Mixed Use" (BMU) (Auckland Council, 2019, p. 13). As granular and legalistic as this sounds, when their representative argued that "there is more than sufficient currently zoned and available BLIZ land for the next few decades", the opposing parties pointed out that the key difference between BLIZ and BMU is that residential development is permitted on the latter and that approximately half of all BMU land in Auckland is used for residential rather than commercial development (p. 21). This accusation was deflected by the representative of the applicants simply telling the panel that he "considers this unlikely because of the ample provision of diverse housing types elsewhere" (p. 18).

Less than a year later, another plan change in the Warkworth area was put forward by five landowners owning 76 ha, seeking to convert the FUZ they inhabited into residential zoned land, bypassing previous planning around the release of land (Auckland Council, 2020). Despite responses from local authorities and planners that this change "goes entirely against the wishes of the community and the Auckland Council within a few months of the Warkworth Structure Plan 2019 being completed" (p. 12), the argument presented by landowners and developers that the delivery of land was the highest priority for the city was accepted (p. 57). Such decisions, as discussed in the previous chapter, increase the speculative value of any other piece of peri-urban land under similar circumstances and encourage the acceleration of fragmentation. Through this process of contesting the interpretation and execution of the Unitary Plan, urban sprawl and the financialization of real estate continue to be enabled.

In Chapter 3, I related how a rise in public concern over the pace of Auckland's urban sprawl led to the creation of a draft National Policy Statement to protect "Highly Productive Land" in 2019 and that it remained a draft for some three years. Over this same period, the Labour-led government wrote and passed the National Policy Statement on Urban Development 2020 (NPS-UD). In order to deal with the country's housing crisis from the supply side alone, the NPS-UD 2020 replaces older policies and sets out a series of positions that all local authorities must take towards development (New Zealand Government, 2020). Most importantly, it demands that:

> Local authority decisions affecting urban environments are responsive to plan changes that would add significantly to development capacity and contribute to well-functioning urban environments, even if the development capacity is unanticipated by RMA planning documents; or out-of-sequence with planned land release.
>
> (p. 12)

This discourse has already been picked up by developers in their plan change submissions and accepted by the environment court. In plan change applications cited earlier that demand the Council meet its obligations to provide land for suburban housing to be built on, the NPS-UD 2020 is the specific justification for prioritizing the release of rural land for development. This policy position, combined with the relations between the interpreters and executors of policy at the level of the territorial field described earlier in this chapter, has created a regulatory environment so poor at refusing development applications that even when it comes to those that are out of step with existing zoning laws, "99.5 per cent of consents get approved", according to a Council planner. Prevailing individual interests and institutional positions ensure any existing measure to protect productive land, public spaces, and ecosystem services is unable to stand up against the unending demand for more capacity.

New Zealand's national policy on urban development saw sudden and bipartisan reform in the later months of 2021. Formulated as a method to increase housing supply in existing urban areas, the alteration to the Resource Management Act of 1991 (which is currently in the process of being reworked in its entirety as of 2022) allows residential lots to host builds of up to three storeys and/or three dwellings without need for an individual resource consent (New Zealand Government, 2021, p. 36), as well as reducing the minimum distance a building must be set back from the property boundary (p. 39). The move has been broadly welcomed as an attempt to remove barriers to urban intensification (Cooke, 2021), but a Rodney Local Board representative described the bill as "yet another blow to efforts to address the yawning gap in infrastructure across Auckland" in Kumeu-Huapai's local newspaper (Pirrie, 2021), comparing the reform to the SHAs in its strategy of attempting to defeat a housing crisis by simply giving private developers more opportunities to build residential housing. They went on to call for more funding of urban infrastructure to support the rapid growth of the city, writing "successive governments led by both major parties have gone for the easy announcements such as the latest one and avoided making the tough calls about addressing the billions of dollars required to provide a liveable city". By encouraging intensification in residential lots that include the formerly rural towns at Auckland's urban periphery, this representative and other officials who work in local authorities (Sachdeva, 2021) or as planners for Auckland Council (OurAuckland, 2021) believe that this amendment will only encourage the pace of residential development at the fringes of the city.

While many other aspects of urban planning have been compromised in the name of providing housing to ease affordability problems, the Unitary Plan and the new NPS-UD are no more successful at levelling off the rise in house prices than SHAs were. Over the past decade, growing cities across the planet have seen soaring prices for housing (Joint Centre for Housing Studies of Harvard University, 2021), fragmentation of productive peri-urban land (Buxton & Butt, 2018, p. 107), and the rapid financialization of housing and property development (Waldron, 2019, p. 687). The unaffordability statistic for Auckland housing cited earlier in this chapter, that the median house cost 11.2 times as the median income, is in fact far worse than the same set of statistics a decade ago. In 2011, this same metric of income to housing cost sat at 6.4 (Performance Urban Planning, 2011). Many other global urban centres have seen a similar near-doubling in housing unaffordability at this time (Performance Urban Planning, 2011, p. 3; Urban Reform Institute and the Frontier Centre for Public Policy, 2022, p. 14) as financial operators across the planet increasingly turn to real estate as one of the few safe bets for a return (Fernandez & Aalbers, 2016). While the provision of new housing is a necessary aspect of providing for the needs of a growing city, it is not a sufficient and total response (Wetzstein, 2021, p. 10), particularly when the forms of urban development presented to meet such a need are entirely subordinated to interests of the group of professionals and landowners that the property market entails. Successive policy efforts have refused to confront the role of housing as a financial asset in the housing affordability crisis. Even as they disregarded a litany of other considerations to make more land available to build on, the land and housing policies detailed above have only encouraged the financial speculation that makes the family home the dominant form of capital accumulation in cities like Auckland.

For the past decade, the ever-more-urgent challenges posed by Auckland's housing crisis combined with a political economic consensus that refuses to stand in the way of financial and development interests that dominate the territorial field of the urban periphery have resulted in a series of policies that place the blame for housing unaffordability on land use regulations and planning provisions. Releasing more land and fast-tracking new builds has failed to solve New Zealand's housing crisis, only serving to enrich financiers, landowners, and developers while worsening urban sprawl. The final section of this chapter looks at the consequences of such a dynamic on work and life for the residents of Auckland in general and of Kumeu-Huapai in particular, as executors of urban development policy reshape rural land at the edge of the growing city to maximize profitability and in doing so limit the possibilities of life there to those that serve the Auckland's FIRE economy.

**The consequences and possibilities of the financialized periphery**

"It's not about land availability, there's plenty of land. We have an infrastructure problem, due to a funding problem, due to a tax base problem". I am sitting in a brand-new café in the brand-new suburb of Hobsonville, a few kilometres east of Kumeu-Huapai, talking to a real estate agent who specializes in the subdivision and sale of future urban land out at the northwest frontier of Auckland. Like many other real estate professionals and landowners in the area, the externalities of their success are starting to concern them. In discussing the consumer market for new suburban housing at the periphery of Auckland City in Chapter 1, I pointed out that city-oriented professionals, often couples who both work full-time and have young children, are the only significant demographic with, as the real estate agent in that section had put it, "the clout" to cover a mortgage for a single-family home in a place like the Huapai Triangle. Without a commensurate effort to supply these very profitable new homes with far less profitable (for the developer) transport and community infrastructure, many of these new residents commute to work and their social lives elsewhere in the city in private vehicles (Stats NZ, 2020a), while local opportunities for working and living have failed to match the unprecedented growth of Kumeu-Huapai that has made the construction and real estate services the biggest earner in town.

Individual agents of the development sector in Auckland's northwestern periphery were often aware in interviews that the possibility of future profitable development in the area had fallen victim to the dramatic success of past ventures. Official documentation on Auckland's future urban land supply states that Kumeu-Huapai should have been ready for development at the end of the 2020s (Auckland Council, 2017, p. 13), yet now that development is ongoing, planners and local officials are worried that creating a structure plan in 2025 to plan for the expected development period of 2028–2032 will only allow more landowners and developers to push their case for being allowed to build residential. As suburban development at the periphery begins to accelerate, locals, planners, and professionals are becoming aware that the formerly rural towns that lie in Auckland's path share a common future as car-dominated exurbs with neither affordability nor sufficient amenities for prospective residents.

Officials and community action groups are demanding that some solution be found, and all run into the same problem. "Tooth fairy's run out of money" an interlocutor from the Rodney Local Board explained, saying the rural Auckland wards of Franklin (south) and Rodney (west and north) don't have the horizontal infrastructure to accommodate new development over the existing expanse of Auckland's urban frontier and that paying for all of this is mostly the responsibility of Auckland Council. Bound by demands from finance (and of course their own interests as a business) to meet a significant rate of profit (Cole, 2017, p. 5; Murphy, 2020, p. 1510), developers only pay for infrastructure and amenities when they absolutely have to. The refrain from Council and from Local Boards, over and over, in interviews, in press releases, and in public meetings, was that there is simply not enough public money to give these new developments the bare minimum of infrastructure. That public coffers should be drier than ever as landowners make unprecedented fortunes is neither a coincidence nor unique to Auckland, though New Zealand has an important aggravating factor. Across the planet, residential housing markets are becoming "the perfect and preferred absorption mechanism" (Wetzstein, 2021, p. 17) for capital holders searching for a safe investment, especially in the volatile economic landscape brought on by the COVID-19 pandemic (Joint Centre for Housing Studies of Harvard University, 2021). The New Zealand government has been unwilling to set a tax on capital gains (most forms of investment other than in real estate are treated as income and taxed as such), making land and housing simply the best possible investments a New Zealander can make, even as the country faces chronic funding issues for healthcare (Stuff, 2021b), education (Stuff, 2021a), and infrastructure. The almost total privatization of the property windfall necessarily creates severe negative impacts on public life in New Zealand.

While the forms of suburban development at Auckland's urban fringe serve mostly to be maximally profitable to the collection of individuals involved in the financing and development of peri-urban land while failing to provide affordable housing supported by the necessary infrastructure to enable work and life, they have also been very successful at transforming the physical, economic, and social landscape of places like Kumeu-Huapai. Reverse sensitivity issues and land fragmentation make productive work harder for small-scale operations in peri urban-Auckland, while the rocketing value of flat land close to the city raises the property rates on the land, pricing agriculture out of operation even if landowners don't simply take the land bankers offer and "go grumbling all the way to bank". These two phenomena – the conversion of productive land into forms only suitable for speculation and the rapid increase of the financial value of the land compared to all other possible uses – work together as part of the same vicious cycle so long as the concerns of the state are aligned with financial success over less lucrative forms of economic and social activity. The financialization of peri-urban land is so crucial to any analysis of its value or "best" use that, as has been described over the three previous chapters, lifestyle block properties are sold advertising the possibilities of a future conversion into suburban zoning, homeowners utilize the speculative value of their house to rapidly trade up to a bigger property, and entire community networks rely on the "legitimated transgression" of an out-of-sequence rezoning of rural land to specifically residential purposes in order for their investments to come to fruition. The inability of local and central government to act against financial interests renders the existence of not only peri-urban horticulture but also any land use that is not maximally profitable – commercial/industrial, community spaces, preserves, and public infrastructure – inimical to the demands of the executors of Auckland's suburban expansion.

If any authority has a desire to mitigate the problems that have been examined here in the case of Kumeu-Huapai, they will have to confront finance capital as an industry and as a set of interests. International policy on land use and housing exists in a political economic context (Gregory, 2009, p. 293), and it must be understood that the think tanks and consultancies that propagate discourses fixated on the global housing crisis as one defined by land supply do so because it benefits the people who pay them (Murphy, 2016, p. 2532). As has been seen over the entire first half of this book, landowners at Auckland's urban fringe, alongside the financial and property sectors, want the best deal they can get with the assets they have and will do all they can to get it. This is to be expected, but allowing these interests to control the future of Auckland's urban development by dominating any discussion of the housing crisis with their own narratives (Squires & White, 2019, p. 171; Wetzstein, 2021, p. 2) is counterproductive to ensuring that people who live and work in the Auckland region have a decent quality of life with access to employment and public services as well as affordable food and housing.

When asked why agents of the central government in particular tended to stand on the side of developers on the housing crisis as an issue of land supply, local officials did not see their counterparts down in Wellington as entirely malicious or incompetent but simply overwhelmed: "they tend to be smashed with work … they skim over the top of stuff". Operating at the broad scale of policymakers at a national level, ministers find that their best opportunity for demonstrating progress on an intractable issue is the announcement of a single measure that may have a positive effect. One member of the local board remarked, not without some irritation, "politicians of a national level, they love to announce things". Chris Penk, speaking in his capacity as an MP of the National Party (now the opposition in parliament) and therefore supposedly of their interests in the area, emphasized "the only routes are persuasion and influence at best, there's no power" when it comes to convincing the rest of the (very much developer-and-landlord-aligned) organization that outwardly presents as the National Party and its affiliates that a change in priorities is necessary. On the other side of the parliamentary aisle, the Labour government's representative for the same region, list MP Marja Lubeck, would have to "go behind the scenes and knock on the doors of the relevant ministers" to convince policymakers of the need to reassess the government's attitude towards urban development and finance. Chris would say in the same interview that the only direct power a minister of parliament possesses over the region they represent is to appoint Justices of the Peace. Even if, as has been detailed in Chapter 1, some of New Zealand political parties have long-standing relationships with the "developer-finance nexus" (Murphy, 2021); the forms that urbanization has taken at the edge of Auckland' sprawl today are as much the result of the inclinations of local landowners and developers seeking to reshape the field to their own benefit than it is the consequence of a failure to create and enforce legislation that would allow less speculatively lucrative land uses to continue to exist.

For this very reason, the central government's strategy towards urban development cannot rely entirely on "market" interests if it truly does seek to provide affordable housing and protect crucial ecosystem services. This particular view has broadly been recently espoused by a number of experts on New Zealand's housing and land use situation (Cole, 2017; Curran-Cournane et al., 2018; Murphy, 2016; Wetzstein, 2021), yet it is worth explicating how such a sentiment relates to the case of Kumeu-Huapai and its relationship to Auckland City. The rising star of Auckland's urban economy, built as it is on state policy that enables the successes of finance and real estate, has bought a corresponding set of conditions that make urban sprawl, land fragmentation, and unaffordable housing nearly inevitable (Kelsey, 2015, p. 87). As will be covered in the following chapter, such a dynamic is not unique to Auckland or New Zealand as a whole. For governments around the world, a confrontation between financial profit-making and the foundational processes that allow these very profits to exist has become an ever more urgent concern (Gregory, 2009, p. 294; Gudeman, 2016, p. 216; Hart, 2000), and it is in the rapidly fragmenting and suburbanizing periphery of growing cities, in places like Kumeu-Huapai, that this contradiction can be clearly seen. Vital land uses are lost to wildly profitable forms of private residential development, and fortunes are made by landowners and development professionals while necessary infrastructure fails to materialize.

If releasing ever more land has failed to provide affordability even as the externalities of such an approach mount ever higher, an alternative strategy must be sought. I am not claiming that places like Kumeu-Huapai, Warkworth and Pukekohe should return to the days where, according to Derry the local historian, one couldn't subdivide land below 50 acres and new developments were only permitted to house "the children's children" of farmers. At the same time, the possibility of land becoming available for development in the future is a powerful upward force on prices (Murphy, 2020, p. 1512). Providing clear boundaries and certainty about future development, as has been proposed under New Zealand's own draft NPS-HPL and in work on the City of Melbourne in Victoria, Australia (Carey et al., 2018), can serve to reduce the impact of financially driven forms of urban sprawl appearing on what was once prime growing land (Buxton & Butt, 2018, p. 141).

Critics of earlier attempts to restrict growth and demand dense urban development were specifically concerned with the high premiums placed on land by regulation and used the work of Glaeser, Gyourko, Grimes, Laing, Lees and others as a theoretical basis for their concerns. Yet, their rhetoric loses its persuasive force in the face of rapid increases in housing affordability wrought by asset inflation and other financial-driven factors that have only been encouraged by the rezoning of peri-urban land for development. Similarly, confronting the unwillingness of the New Zealand government to effectively tax housing as a financial asset offers the possibility of simultaneously addressing the runaway speculative value of residential development and curtailing the privatization of the profits of said development (Wetzstein, 2021, p. 13). For the same reasons, the past four decades of financialization of the productive economy have relied on the alignment of the state with financial interests, and any genuine restriction from the former onto the latter would help destabilize the vicious cycle that has rendered the urban periphery a fallow land bank.

As has been detailed throughout this chapter, however, the urban periphery is not a simple reflection of maps drawn up by policymakers in Auckland or in Wellington, but a field of its own, subject to the resources and social relations of individual agents and organizations seeking to make their fortunes in an urbanizing and financializing world. Indeed, it was when I was looking at a map of Auckland's

Future Urban Zones (Figure 4.4) that a Council planner told me that "the battle for Pukekohe is lost, the FUZ will never be reduced". That now-foundering horticultural powerhouse at the southern frontier of Auckland City, the loss of which has been forecast to raise vegetable prices across the nation by 50 per cent (Deloitte, 2019), was assumed by the planner to soon be lost to suburban development even if current legislation claimed otherwise. They then pointed to another rural town slightly further out from Auckland, Patumahoe, suggesting this was next in line for more of the plan change submissions described in this and previous chapters. Sure enough, a plan change has been lodged and is ongoing (Auckland Council, 2021b). For planners and policymakers in Auckland and around the world, contending with the unending demand for land to be released at the periphery of cities can seem futile, given the financial incentives that drive banks, land developers, landowners, housebuilders, and even individual homeowners to seek a legitimated transgression of any attempt to stand in the way of capital accumulation.

Yet, it is at this moment of seeming impossibility to construct an alternate approach to land use governance that attempting to constrain the apparently unstoppable force of finance capital in land and housing has never been more acutely necessary. As our cities grow faster than ever before, as our global climate rapidly destabilizes, and as economic inequality reaches new heights, understanding how these challenges intersect only becomes more important. For this reason, while Kumeu-Huapai remains the place in which I have conducted fieldwork and will chiefly refer to for substantive information about financially driven urban sprawl, a complementary perspective is needed. The following chapter takes a comparative look at two formerly rural towns at the periphery of Anglo-colonial urban centres that today find rapid suburban growth to be impinging on their regional foodbowl and constraining ecosystem services. Gisborne, north of Melbourne in Victoria, Australia, and Fresno in California, USA, both face a similar set of challenges to those seen in places like Kumeu-Huapai but bear useful points of contrast in terms of scale, governance, and approaches to urbanization, as well as provide insights into how climate change will interact with the fragmenting and financialized urban periphery.

## Bibliography


Auckland Council. (2017). *Auckland Future Urban Land Supply Strategy July 2017.* Retrieved September 14 from https://www.aucklandcouncil.govt.nz/plans-projects-policies-reports-bylaws/our-plans-strategies/topic-based-plans-strategies/housing-plans/Documents/future-urban-land-supply-strategy.pdf

Auckland Council. (2019). *Decision Following the Hearing of a Plan Modification (PPC25 – Warkworth North) to the Auckland Unitary Plan under the Resource Management Act 1991.* Retrieved from https://www.aucklandcouncil.govt.nz/UnitaryPlanDocuments/pc-25-decision.pdf

Auckland Council. (2020). *Decision Following the Hearing of a Plan Modification (Private Plan Change 40: Warkworth-Clayden Road) to the Auckland Unitary Plan under the Resource Management Act 1991.* Retrieved from https://www.aucklandcouncil.govt.nz/HearingDocuments/pc40-dec-2020-10-05.pdf

Auckland Council. (2021a). *Auckland Unitary Plan Operative in Part*. Retrieved September 10 from https://unitaryplan.aucklandcouncil.govt.nz/pages/plan/Book.aspx?exhibit=AucklandUnitaryPlan_Print

Auckland Council. (2021b). *Auckland Unitary Plan Operative in Part Proposed Plan Change 55(Private): Patumahoe South Summary of Decisions Requested*. Retrieved from https://www.aucklandcouncil.govt.nz/UnitaryPlanDocuments/pc-55-sdr.pdf

Auckland Council. (2021c). *Draft Proposed Variation 1 to Proposed Plan Change 5 – Whenuapai Draft Section 32 Report*. Retrieved April 19 from https://www.aucklandcouncil.govt.nz/plans-projects-policies-reports-bylaws/our-plans-strategies/unitary-plan/auckland-unitary-plan-modifications/Pages/details.aspx?UnitaryPlanId=105

Barker and Associates Urban and Environmental. (2020). *S32 Assessment Report Drury East Private Plan Change Request*. A. Council. Retrieved May from https://www.aucklandcouncil.govt.nz/UnitaryPlanDocuments/pc-49-section-32-report.pdf

Bourdieu, P. (2005). *The social structures of the economy*. Cambridge: Polity.

Brown & Company Planning Group. (2015). *Huapai Triangle – Proposed Variation to the Proposed Auckland Unitary Plan to enable the Huapai Triangle Special Housing Area* [Provisional planning document handed out to local community members and development interests].

Buxton, M., & Butt, A. (2018). *The future of the fringe: The crisis in Peri-urban planning*. Melbourne: CSIRO Publishing.

Carey, R., Sheridan, J., & Larsen, K. (2018). *Food for thought: Challenges and opportunities for farming in Melbourne's foodbowl*. The University of Melbourne. Retrieved August 8 2021 from https://science.unimelb.edu.au/foodprint-melbourne/publications/report-challenges-and-opportunities-for-farming-in-melbournes-foodbowl

Cole, V. (2017). Why landlords and investors love the Unitary Plan, and why you shouldn't. *Economic and Social Research Aotearoa, 1*, 1–13.

Cooke, H. (2021). *Sweeping Townhouse Bill Passes into Law with Support of Labour and National, will Force Councils to Allow More homes*. Retrieved 24 March 2022, from https://www.stuff.co.nz/national/politics/300478047/sweeping-townhouse-bill-passes-into-law-with-support-of-labour-and-national-will-force-councils-to-allow-more-homes

Curran-Cournane, F., Golubiewski, N., & Buckthought, L. (2018). The odds appear stacked against versatile land: Can we change them? *New Zealand Journal of Agricultural Research, 61*(3), 315–326.

de Roiste, M., Fasianos, A., Kirkby, R., & Yao, F. (2021). Are housing wealth effects asymmetric in booms and busts? *The Journal of Real Estate Finance and Economics, 62*(4), 578–628.

Deloitte. (2019). *New Zealand's Food Story: The Pukekohe Hub*. Retrieved from https://www2.deloitte.com/nz/en/pages/primary/articles/case-study-pukekohe-hub.html

Diagne, A. F., Kurban, H., & Schmutz, B. (2018). Are inclusionary housing programs colorblind? The case of Montgomery County MPDU program. *Journal of Housing Economics, 40*, 6–24.

Fernandez, M. A., Sánchez, G. E., & Bucaram, S. (2021). Price effects of the special housing areas in Auckland. *New Zealand Economic Papers, 55*(1), 141–154.

Fernandez, R., & Aalbers, M. B. (2016). Financialization and housing: Between globalization and varieties of capitalism. *Competition & Change, 20*(2), 71–88.

Glaeser, E. L., & Gyourko, J. (2002). Zoning's steep price. *Regulation, 25*, 24.

Glaeser, E. L., & Gyourko, J. (2003). The impact of building restrictions on housing. Affordability," *Economic Policy Review, 9*(2), 1–19.

Glaeser, E. L., Gyourko, J., & Saks, R. (2005). Why is Manhattan so expensive? Regulation and the rise in housing prices. *The Journal of Law and Economics*, *48*(2), 331–369.

Google LLC. (2022). *Google Satellite via QGIS*. Retrieved from http://mt0.google.com/vt/lyrs=s&hl=en&x={x}&y={y}&z={z}. Source: http://mt0.google.com/vt/lyrs%3Ds%26hl%23Den%26x%23D%27Bx%27D%26y%23D%27By%27D%26z%23D%27Bz%27D&zmax=18&zmin=20. Provider: wms.

Gregory, C. (2009). Whatever happened to economic anthropology? *The Australian Journal of Anthropology*, *20*, 285–230.

Grimes, A., & Liang, Y. (2009). Spatial determinants of land prices: Does Auckland's metropolitan urban limit have an effect? *Applied Spatial Analysis and Policy*, *2*(1), 23–45.

Gudeman, S. (2016). *Anthropology and economy*. Cambridge: Cambridge University Press.

Hart, K. (2000). Money in an Unequal World. *Anthropological Theory*, *1*(3), 307–330.

Hong, L. (2020). *The Need to Build: The Demographic Drivers of Housing Demand*. The New Zealand Initiative. https://www.nzinitiative.org.nz/reports-and-media/reports/the-need-to-build/document/684

Joint Centre for Housing Studies of Harvard University. (2021). *Home Prices Are Skyrocketing in Most Markets*. Retrieved September 13 from https://www.jchs.harvard.edu/son-2021-home-prices

Kelsey, J. (2015). *The FIRE economy: New Zealand's reckoning*. Wellington: Bridget Williams Books.

Lees, K. (2019). Quantifying the costs of land use regulation: Evidence from New Zealand. *New Zealand Economic Papers*, *53*(3), 245–269.

Manville, M., Lens, M., & Monkkonen, P. (2022). Zoning and affordability: A reply to Rodríguez-Pose and Storper. *Urban Studies*, *59*(1), 36–58.

Martin, S. L., & Norman, D. S. (2020). *An Evidence-Based Approach Does the Rural Urban Boundary Impose a Price Premium on Land Inside It? Final Report* (Chief Economist Unit, Issue). Retrieved from https://www.aucklandcouncil.govt.nz/about-auckland-council/business-in-auckland/Reports/does-the-rub-impose-a-price-premium-on-land-inside-it-20-Feb-2020.pdf

McLeay, C. (2020). Housing Accords and Special Housing Areas Act 2013 and the erosion of democracy. *New Zealand Geographer*, *76*(2), 127–134.

Murphy, L. (2016). The politics of land supply and affordable housing: Auckland's Housing Accord and Special Housing Areas. *Urban Studies*, *53*(12), 2530–2547.

Murphy, L. (2020). Performing calculative practices: Residual valuation, the residential development process and affordable housing. *Housing Studies*, *35*(9), 1501–1517.

Murphy, L. (2021). *New housing supply, housing affordability and the developer-financier nexus*. Auckland: NZGS Auckland Dialogues.

Murray, C. K. (2021). Marginal and average prices of land lots should not be equal: A critique of Glaeser and Gyourko's method for identifying residential price effects of town planning regulations. *Environment and Planning A: Economy and Space*, *53*(1), 191–209.

New Zealand Government. (2020). *National Policy Statement on Urban Development 2020*. Retrieved from https://www.hud.govt.nz/urban-development/national-policy-statement-on-urban-development-nps-ud/

New Zealand Government. (2021). *Resource Management (Enabling Housing Supply and Other Matters) Amendment Act 2021*. Retrieved from https://www.legislation.govt.nz/act/public/2021/0059/latest/LMS566049.html#LMS566031

O'Flaherty, B. (2003). Commentary (on Glaeser and Gyourko). *Federal Reserve Bank of New York Policy Review*, *9*(2), 41–43.

OurAuckland. (2021). *Auckland Council Calls for Changes to Government's Enabling Housing Supply Bill*. Auckland Council. Retrieved 24 March from https://ourauckland.aucklandcouncil.govt.nz/news/2021/11/auckland-council-calls-for-changes-to-government-s-enabling-housing-supply-bill/

Performance Urban Planning. (2011). *7th Annual Demographia International Housing Affordability Survey: 2011 Ratings for Metropolitan Markets*. Retrieved from http://www.demographia.com/dhi2011.pdf

Pirrie, P. (2021). Phelan's Local Board update. *Kumeu Courier*, November 2021, p. 45.

Sachdeva, S. (2021). *The Intensifying Debate over Housing Reform*. Newsroom. Retrieved 24 March from https://www.newsroom.co.nz/the-intensifying-debate-over-housing-reform

Squires, G., & White, I. (2019). Resilience and housing markets: Who is it really for? *Land Use Policy*, *81*, 167–174.

Stats NZ. (2020a). *Place Summaries: Kumeu-Huapai*. Stats NZ. Retrieved June 4 from https://www.stats.govt.nz/tools/2018-census-place-summaries/kumeu-huapai

Stats NZ. (2020b). *Which Industries Contributed to New Zealand's GDP?* Retrieved June 7 from https://www.stats.govt.nz/tools/which-industries-contributed-to-new-zealands-gdp

Stuff. (2021a). *Budget 2021: More Money Needed Across All Aspects of Education*. Retrieved September 14 from https://www.stuff.co.nz/timaru-herald/news/125150386/budget-2021-more-money-needed-across-all-aspects-of-education

Stuff. (2021b). *Doctor Shortage Forcing GP Clinics to Turn away New Patients*. Retrieved September 14 from https://www.stuff.co.nz/national/health/124477572/doctor-shortage-forcing-gp-clinics-to-turn-away-new-patients

The Environment Court at Auckland. (2020). *Decision No. [2020] NZEnvC 214 in the Matter of the Resource Management Act 1991 and the Local Government (Auckland Transitional Provisions) Act 2010 and of an Appeal under Section 156(1) of the Local Government (Auckland Transitional Provisions) Act 2010 Between Self Family Trust (ENV-2016-AKL-000199), Apellant, and J & F Gook, Section 274 Party, and Auckland Council, Respondent, and Ted Edwards, Section 274 Party.* Retrieved from https://environmentcourt.govt.nz/assets/Documents/Publications/2020-NZEnvC-214-Self-Family-Trust-v-Auckland-Council.pdf

Urban Reform Institute and the Frontier Centre for Public Policy. (2022). *Demographia International Housing Affordability 2022 Edition*. Retrieved 24 March from http://www.demographia.com/dhi.pdf

Waldron, R. (2019). Financialization, urban governance and the planning system: Utilizing 'development viability' as a policy narrative for the liberalization of Ireland's post-crash planning system. *International Journal of Urban and Regional Research*, *43*(4), 685–704.

Wetzstein, S. (2021). Toward affordable cities? Critically exploring the market-based housing supply policy proposition. *Housing Policy Debate*, *32*, 1–27.

WSP New Zealand Limited. (2022). *MGD Foundation Course 2022*. Retrieved from https://nzetc.wsp.com/w/courses/341-mgd-foundation-course-2022

# Part Two

# Life at the edge of the city

# 5 National and international comparative case studies on financially driven peri-urban development

For many of the older professionals I interviewed, those who had once worked in Auckland's northwest in horticulture before taking advantage of New Zealand's economic realignment to find a more lucrative career in government, development, or both, their time in Kumeu-Huapai was coming to an end. "A lot of people I know ... are moving north or prepping to do so" one such official told me. Sitting on multiple hectares of land in places like Kumeu-Huapai and nearby Waimauku or residing in beach houses overlooking the ever-busier Muriwai coast, these long-standing professionals on all sides of the debate of the suburbanization of Kumeu-Huapai were thinking about how their massively appreciated property assets would guarantee their retirement plans. One interlocutor had in fact already retired, though they maintained a position in a local community action group advocating for development forms guided by the needs of the community. As I walked through the villa in which they had settled down, I commented on how impressive the place was, to which they replied with a chuckle "this is what selling fifteen hectares gets you".

Amid dramatic (and probably accurate) forecasts that the area would soon be indistinguishable from any other sprawling suburb of Auckland, people who had once had the money to seek out "lifestyle block paradise" in Kumeu-Huapai were cashing in their land and moving on to greener pastures. The question of where said pastures reside is not easily answered. As one real estate agent put it, "I moved here because I was sick of the traffic in Auckland, I don't know where I'll move to once I get sick of the traffic here". Some, as the official had implied, simply step a few kilometres further away from Auckland City centre while still remaining within driving distance, ensuring that fragmentation of rural land continues to be very profitable at the periphery. Others, however, especially those stepping away from work to retire permanently, move to bucolic rural communities or the peripheries of other towns across the country. Along with these wealthy customers has come the "financier-developer nexus" that has transformed Auckland's own urban periphery from a uniquely productive landscape into a speculative investment vehicle. Across every corner of Aotearoa, the FIRE economy that has so entirely subsumed political and social life in its urban centres can be ever more clearly seen reshaping work and life.

This chapter seeks to place the observations made about Kumeu-Huapai in the first half of this book in context with how this historical process of financialized development has been expressed at three scales. The first is at the regional, by looking

DOI: 10.4324/9781003400325-8

at how Kumeu-Huapai is reflective of the rest of Auckland's urban periphery and also at the ways in which it is a unique case. Next, at how the financialization of real estate in small towns across New Zealand, at distances from urban centres untenable for even the most determined city-oriented commuter, is occurring as housing unaffordability in Auckland continues to escalate. Finally, much of the latter part of this chapter is devoted to a comparative look at suburban development at the periphery of two very similar yet much larger colonial cities that are caught between being foodbowls and financial centres. In examining the case of Gisborne, a formerly rural town, 45 minutes' drive from Melbourne in Victoria, Australia, and the setting for the 18th and most recent season of competitive home renovation show *The Block*, I discuss how – as at the end of Chapter 2 – the unending demands of peri-urban land to be converted to its most financially lucrative uses run into barriers for further accumulation created by their own success. Then, by looking at the relationships between governance and the dominant economic sectors of the deeply unequal city of Fresno in California, USA, I expand on the point in the previous chapter – that Auckland's crises of housing unaffordability and suburban sprawl are enabled by individuals who benefit from them – to argue that the broader planetary phenomena of rapid urbanization and financialization at the urban periphery exist as a transfer of wealth from the working populace to financial centres and to a small class of professionals who make their living at the frontier of speculative expansion. By following this expanding focus of cases – from regional context, to national consequences, to two international comparative studies – I intend to show how the dynamic shown in Kumeu-Huapai in earlier chapters possesses an important explanatory power for challenges faced by an urbanizing humanity throughout Aotearoa and the world.

## "We're really the last frontier" commonalities and disparities across peri-urban Auckland

One very dramatic heuristic for the pace of development in Auckland, both in my own experiences of fieldwork and in the daily lives of interlocutors, was the quantity of vehicle traffic. As mentioned previously throughout this book, local ministers spent much of their office time responding to complaints from residents about their commute, real estate agents worried they were losing buyers by having their experience of rural paradise soured by standstill traffic, and developers demanded the state summon whatever resources remained at its disposal to increase road capacity and enable further suburban development. Yet, while some would make a point of saying the traffic situation in Kumeu-Huapai was uniquely dire, others pointed to massive peak-hour snarls in the other new frontiers of Auckland – Warkworth and Silverdale to the north, and Drury and Pukekohe to the south. Discussing a political fight over funding for a highway expansion in Silverdale, a planner told me "that's just the same as here", before noting how similar promises made to the residents of Kumeu-Huapai to allow development had also failed to materialize.

From at least a cursory visual standpoint as a resident of the city, and from conversations with long-time Kumeu-Huapai residents, the spread of very profitable forms of suburban development is if anything more prevalent to the north and south than to the west. The Huapai Triangle may have been the largest SHA to be created, with approximately 1,200 potential dwellings, but recent developments such as Milldale to the north of Auckland (Milldale, 2022) and the newly approved rezoning of East Dury to the south (Forbes, 2022) are each more than 300 hectares in area and promise more than 4,000 houses apiece. Developments of such a scale, literally approaching the footprints of smaller New Zealand cities such as Napier and Nelson, dominate the skylines and the headlines of Auckland's urban periphery. These massive new dormitory suburbs of Auckland City were often brought up to me by developers or officials as object lessons in how big developers worked in the region. More than once, Milldale was used as an example of how developers could sidestep zoning and infrastructure requirements from the beleaguered Council by creating private infrastructure agreements for the new homeowners to bear infrastructure costs (Milldale Infrastructure LP, 2020), while the billion-dollar legal battles and even criminal trials (Kilgallon, 2021) over land in Pukekohe, Pokeno, and Drury were seen by some interlocutors as informative and very entertaining case studies of land bankers and other agents "ripping each other off" in their desperate bids to make a fortune at the urban periphery. While Fletcher and Cabra continue to plan for new phases of development around Kumeu-Huapai, it sometimes seemed, as I continued to interview and observe, that if anything many of the real, emblematic, "important" instances of financialized sprawl were happening elsewhere.

Part of the reason why the northern and southern fringes of the city tended to dominate conceptions of sprawl in media and in the minds of some interlocutors is the sheer amount of land at these two frontiers that has been urbanized in living memory, particularly from the vantage point of anyone driving across the Auckland region. Between 1970 and today (see Figure 5.1), proximate rural areas such as Albany to the north and Botany to the south (both common examples given by older interviewees) were entirely incorporated into urban Auckland, while rural towns that had once operated on mostly separate social and economic lives from the city centre, like Papakura and Orewa, became exurban settlements directly oriented towards the functioning of Auckland City.

As detailed in Chapter 2, an identical historical process transformed peri-urban horticultural areas in the west such as Te Atatu and Henderson into the newest extent of Auckland's urban core and led to a reorientation of work and life in the more distant Kumeu-Huapai, but a key difference in perceptions between the cases is the sheer continuous distance of development along the north-to-south axis. To describe this concretely, if you were driving from north to south across New Zealand via State Highway 1, you would pass over the boundary line of Auckland region just north of the small town of Te Hana, 90 km by road from the city centre, then come 40 kilometres more driving, through the northern service town of Wellsford, conservation areas, and the larger farms, estates, and lifestyle blocks owned by people who usually have no need to travel to Auckland proper every day, though I have met several people who live this far from the city and still work and socialize there, making their commute time more than three hours a day at a minimum. On this imaginary trip, once you pass through the Johnstone Hills Tunnel, the view of hills on

*Figure 5.1* Aerial photography of the formerly peri-urban settlements of Albany (top) and Botany (bottom) in 1970/72 and 2020, comparing the lifestyle blocks and exurbs of the former with the general urbanization of the latter.

*Figure 5.1* Continued

the left-hand side gives way to the former beach towns of Orewa, Red Beach, and Whangaparāoa. As you pass through the rapidly growing exurb of Silverdale (of which Milldale is a significant new addition to its western frontier), you have one final interlude of lifestyle blocks for some 10 kilometres, before finding yourself in Albany and the North Shore. From this point, the concept of being within Auckland is not some abstract theoretical concept about social relations towards land, but an unavoidable visual ideology impressed into the landscape by decades of suburban expansion. Driving south through on a state highway flanked by industrial parks and dense clusters of single-family homes, you will spend 50 entire kilometres passing through what can be generally glossed as Auckland City, with the only structural differentiation being a brief interlude of high-rise office buildings of the CBD dominated the left side of your view as you crest the harbour bridge. The highway takes you through once-distinct settlements of Ellerslie, Manukau, and Papakura, all conjoined into this single, very specific urban form. Yet, even once you have driven past this mass, just as Silverdale and Kumeu-Huapai act as new outposts of the city to the north and west, you will see, emerging from the hills like mushrooms after rain, new suburbs being built in Pokeno (technically just past the boundary of Auckland region), in Bombay, all along the Highway. More than once, I would hear a joking refrain from interviewees, along the lines of "It takes me longer to get out of the city every year, and I'm not just talking about the traffic!".

The second reason for Kumeu-Huapai being a less prominent feature of discussions about urban sprawl than other sites of fragmentation and suburbanization in Auckland is that, at least as of 2021 during my fieldwork, Auckland's southern frontier could still offer an argument against development due to their undeniable agricultural productivity. As I have referred to in previous chapters, the southern town of Pukekohe (a few kilometres west of State Highway 1 and therefore not visible on our highway trip across the Auckland Region) is facing the same vicious cycle of financialization of land uses and increasing difficulties for smaller horticultural producers that reduced the possibilities of Kumeu-Huapai's future to that of land bank and dormitory suburb, only some 30 years later and 30 kilometres further out from the city centre. Even as the "bonanza-like conditions" of the financializing periphery (Wetzstein, 2018, p. 1) push out productive land uses and thus justify further development, this location, comprising only 9 per cent of New Zealand's vegetable production industry in land area, continues to produce 26 per cent of the (dollar) value of the nation's horticulture industry, and 81 per cent of produce grown in the Pukekohe "hub" goes straight to the local consumer market rather than to export (Deloitte, 2019, p. 20). The loss of the area to development has been forecast to raise the price of vegetables across the country by around 50 per cent (p. 9). Some residents of Kumeu-Huapai may remember a time when the town's horticultural production was "the biggest in Auckland", but that title undoubtedly sits with the Pukekohe today.

As public expressions of concern from environmental scientists, planners, and residents over the acceleration of Auckland's outward expansion began to grow in the late 2010s, it was therefore Pukekohe that became the clearest contemporary case study for how the urban periphery was becoming subject to abstract financial demands for profitability over less lucrative but crucially important functions like growing the food we eat. Articles in *Radio New Zealand* showed Pukekohe horticulturalists either facing difficulties or being forced to shut down operations entirely (Corlett, 2018; Radio New Zealand, 2019, 2020) while being offered sums of up to a million dollars per hectare for their land by land bankers hoping to, in turn, be bought out by developers. In 2019, the minister for the environment made reference to Pukekohe in his announcement of the draft National Policy Statement on Highly Productive Land, warning:

> Every day, truckloads of broccoli and other winter greens leave from the Auckland region, not just to feed Auckland, but also to feed the South Island ... If we lose these soils, the price of vegetables will go up and we'll be a less healthy country.
>
> (Dreaver, 2019)

While today Kumeu-Huapai's role as a productive centre for Auckland must be pieced together from records and the testimony of those who remember, the Pukekohe case allows us to observe how the financialization of a productive periphery by individual agents and by policy occurs in real time.

Conversely then, the much closer and more fragmented Kumeu-Huapai offers the chance to see how the endless demands for maximal profitability for peri-urban land transform life at the frontier of the growing city at a far more abstracted (or "advanced") level than seen anywhere else at the periphery. While every one of the new "nodes" of urban Auckland has been growing over the past decade, the expansion of Kumeu-Huapai in particular is not only unprecedented in the history of the town but among any of Auckland's other suburbs in waiting today (see Figure 5.2). In the coming years, the massive developments being drawn out to the north and south of the city will make their own impact. Yet, at the time of my fieldwork, Cabra's past and current projects in Kumeu-Huapai, along with the

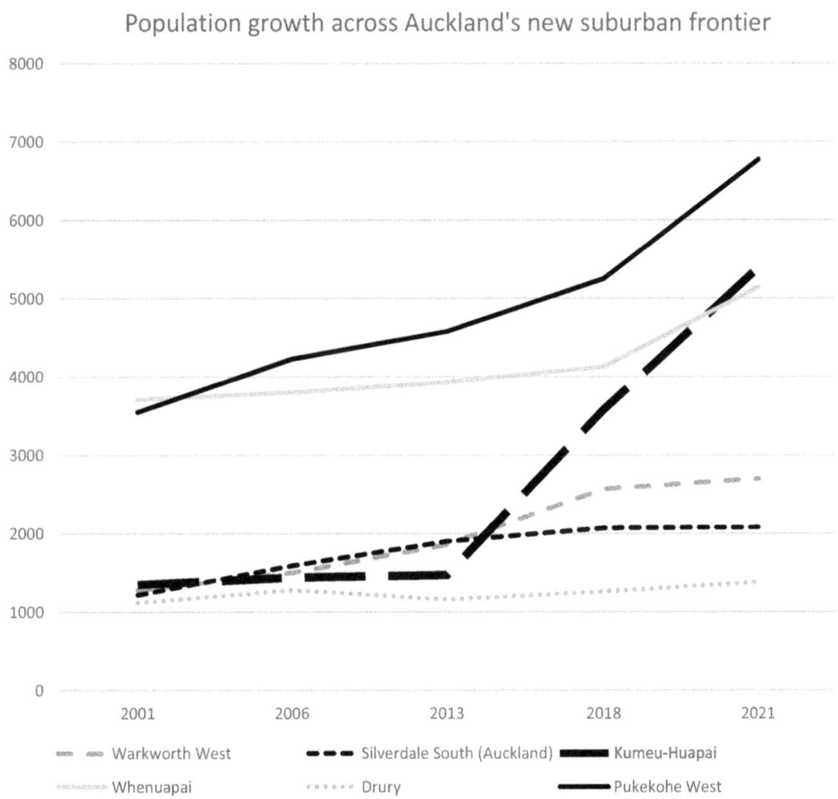


*Figure 5.2* Population over time in notable census areas corresponding to Warkworth, Silverdale, Kumeu-Huapai, Drury, and Pukekohe ... Kumeu-Huapai is indicated by the thick dashed line. Note the jump in population growth compared to other areas (Stats NZ, 2022).

purchases of land nearby by Fletcher Residential, are the most visible expression of how financialized approaches to the periphery reorient work and life towards the city centre once small-scale horticulture has been driven out.

Today, farmers in Pukekohe are arguing that maximally profitable forms of suburban development must be curtailed through binding national-level policy and that the consequences for failing to do so are severe and irreversible. "What the message really is that we've got to get across to people, is if you have a look at history and see what was actually cropping, from north coming down, all of it's gone" a grower was reported telling Radio New Zealand, before saying "we're really the last frontier" (Corlett, 2018). One interlocutor, a worker at one of the two extant horticulture operations in Kumeu-Huapai, has several colleagues working in Pukekohe and noted "down in Pukekohe they used to be pretty confident, because they were number one, but now of course the small ones are going, then the medium ones". In Chapter 1, the situation in Pukekohe today was used to make inferences about the generalized collapse of horticulture in Auckland's northwest 30 years ago, and this comparison is just as helpful for the future of Pukekohe's land use as for the past of Kumeu-Huapai's. Despite the challenges they faced in their work, the labour that horticulturalists put into the land in Auckland's northwest combined with its proximity to transport methods to the city made Kumeu-Huapai uniquely valuable, marked on maps in the 1970s and 1980s as growing land of the highest quality. Today, the landscape has been successfully transformed in discourses and in quantitative measurements into a fallow land bank, of best use as an investment and as a place for developers to "build fast, sell high and leave quick". If the courts and the state as a whole have decided that horticultural operations must be competitive with the most lucrative alternative land use regardless of any measurement relating to Land Use Classification, then this remaking of the past and of possible futures will make its way to Pukekohe as well. Should this particular formation of state and individual interests persist, then perhaps in 20 years' time a researcher studying Auckland's urban periphery will meet former horticulturalists who respond bitterly to their questions about whether this fast-growing southern suburb of Auckland had once been horticulturally valuable.

## Priced out: real estate investor attitudes towards small towns across Aotearoa

Kaikohe is a small town of less than 5,000 people in the far north of New Zealand's North Island, some 240 km north of Auckland City by car. Just take State Highway 1 and then a left turn onto State Highway 12. Kaikohe is situated in such a way as to be roughly equidistant from the Northland west coast beach towns Opononi and Hokianga and the (much wealthier) east coast holiday destinations of Paihia and Russell. As of 2018, a third of the population aged over 15 worked full time while another 40 per cent were not in labour force at all (Stats NZ, 2018). The median annual income was $19,000 NZD. By 2021, residents were finding themselves subject to a rapid rise in attention from house buyers, attracted to the town by the simple fact homes are still sold there for under $300,000 NZD (Olley, 2021). As property values increased by a quarter over a single year and landlords averaged an annual profit margin of almost 8 per cent, residents found it suddenly much harder to find a house to rent or buy. One long-time real estate agent in the area was shocked by the rapid change in houses and land were being valued, saying of one family of tenants who found they had to leave after their landlord sold up:

> These people and all their children, where are they going to go? These are the worst things for families. You take that into consideration that the kids may have to change schools, they may not even find anywhere. We haven't really got much in the way of emergency housing here. There's only a couple of motels.

Another real estate professional interviewed by Radio New Zealand pointed to the massive cost of houses in the Auckland region leading to a spike in buyers seeking relatively cheaper real estate anywhere they could. With its subtropical climate, relative proximity to Auckland and to several renowned beach destinations, and indeed, its relative poverty compared to the wealth held by property owners in Aotearoa's largest city, the Northland region has been a common destination for prospective buyers in recent years. Reports abound of Aucklanders buying houses around Whangarei, the region's largest city, with large cash offers "sight unseen" (Kuprienko, 2022):

> We've had some astronomical prices driven by the competition from out-of-town buyers. It's people from Wellington and Auckland fighting … The community here historically was a chill and very down to Earth place but because it's become more attractive to Aucklanders, we are noticing a bit of change. You can tell by the kinds of cars driving around and the level of social-economic status.

These wealthy buyers are also particularly interested in any empty sections, seeking to build their dream home on the coast with the cash proceeds from selling up in Auckland. More than simply transforming its urban periphery into an investment vehicle, the realignment of Auckland's dominant economic interests towards abstracted returns from financial assets provides the opportunity for individuals to acquire much less "valuable" homes and land further afield, extending the financialization of rural land well beyond the urban periphery.

Kaikohe is, in fact, one of ten separate towns listed in a 2021 article in the business section of the news site *The Spinoff* as "places where you can still buy a house for under $300K" (Andrew, 2021). Kaikohe was the only candidate in Northland, but other notable entries in uniquely affordable real estate opportunities for the canny investor included towns on the South Island's West Coast and Southland, along with rural corners of Taranaki and Hawke's Bay in the North Island (see Figure 5.3). Unsurprisingly, property values in these areas have risen dramatically in the last few years, as the scenes of massive financial accumulation and tenant dislocation previously described in Kaikohe become relevant across Aotearoa. The districts of Buller and Westport, home to several of the west coast towns listed in the article, have seen price rises of over 30 per cent in a single year, at the same time as widely reported incidents of flooding and infrastructure failure in the area (Naish, 2022). The very real concerns for the integrity of the houses and viability of land for development seem to be, as explained by the property valuer in Chapter 3 "priced in". Just as in peri-urban Whangarei and across Northland, buyers are turning up and purchasing properties within minutes of release with "fistfuls of cash" (Smith, 2022) with similarly destabilizing results for the people who live and work in these towns.

The extension of the "territorial field" of wildly profitable forms of suburban development, with all the implications sketched out in previous chapters, follows close behind such a dramatic rise in demand from wealthy buyers. Interlocutors tended to find these new suburban developments far from Auckland to be particularly noteworthy and indicative of some kind of radical transformation of the landscape, whether they heard of them in their own professional contexts or from

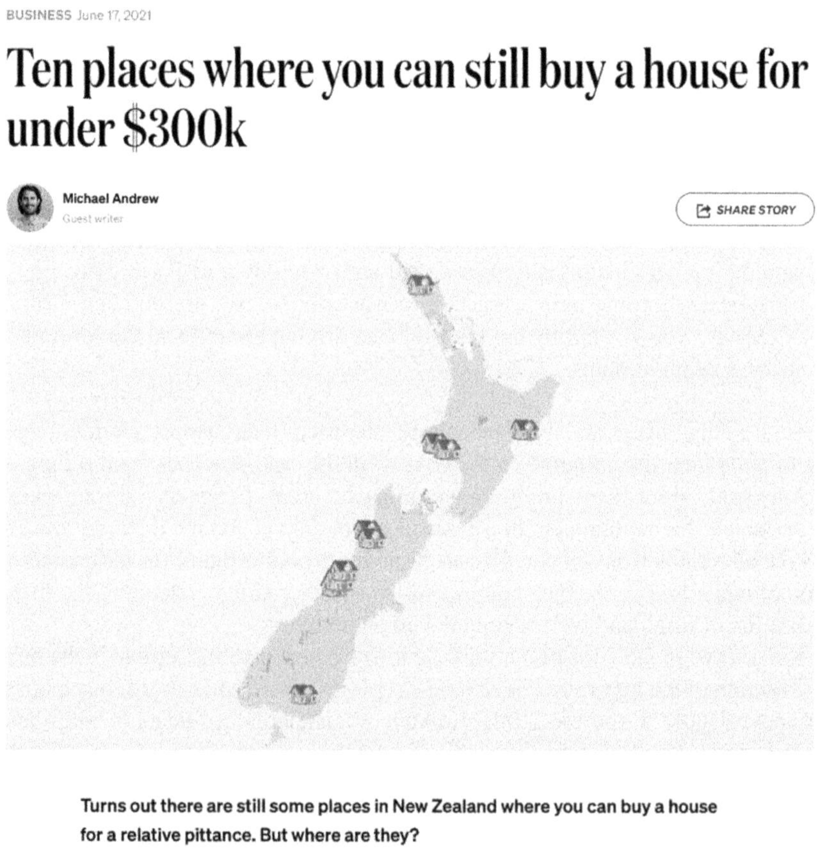


*Figure 5.3* The header of Michael Andrew's article in *The Spinoff* detailing the remaining places in the country where a house on freehold land could still be purchased for below $300,000NZD as of 2021 (Andrew, 2021).

the media. In the Heretaunga plains of the Hawke's Bay region, one of the places buoyed by rapid international freight, industrial production, and immigrant labour to overtake Auckland's own horticulture producers, experts gather to attempt to find some kind of balance between the sudden rise in demand for land in the area to be released for suburban development and the productive value of the land as it is today (Jeffares, 2022). Residents in Rotorua, a small city and tourist destination some 240 km by road south of Auckland, have recently discovered that their council has been conducting closed-door negotiations to sell off ten of its parks and nature reserves for the purpose of development into suburban housing (Desmarais, 2022). Even in places well removed physically from Auckland's urban economy, the turn towards financial and meta-financial values for land and housing amid the urbanization of the population is an increasingly important phenomenon.

For this reason, the particular arrangement of interests and confluence of historical factors that transformed Kumeu-Huapai within living memory can be used to understand the broader dynamic of rapidly accelerating financialized urbanization at a regional and even national scale. The usefulness of this comparison does not end with an understanding of the interaction between central government and development interests and the shift in ruling priorities from production to finance. While I spent part of Chapter 2 arguing against a single and supposedly linear progression from agrarian colonial origins to unavoidable suburbanization, the process of a vicious cycle of fragmentation and speculation followed by maximally profitable suburban development without concurrent investment in the necessary infrastructure to support human life that is so dramatically visible in narratives of Kumeu-Huapai's development can also be seen at different "stages" of its life cycle across Aotearoa, from Kaikohe to Southland. Individual agents at these new peripheries not only share common approaches and motivations to those in Auckland, they are quite often literally the same people. A lawyer who has spent decades representing agents in property development said in one interview "The Cabras, the Hoppers, they have a history of buying land at the edge of rural and coastal settlements and waiting", telling me that the Hoppers (who are building the luxury retirement village on Cabra's Huapai Triangle) rezoned and suburbanized the entirety of Pauanui, now one of New Zealand's more affluent resort towns, using the capital gained from this process to begin their work across the Auckland region in earnest some 40 years ago. "That's where they started".

Because of this close relationship between development at Auckland's urban periphery and in rural areas across the country, the past four decades of horticultural collapse and rising financial interest in suburbanization in Kumeu-Huapai offer a particularly clear depiction of how these same agents work elsewhere. At the time of our interview, initially held to discuss their past work in northwest Auckland, this same lawyer was working on a plan for a rezoning application in New Plymouth, the regional capital of the very affordable North Island region of Taranaki. They told me that this range of development work for landowners across the nation to, as a local official in the previous chapter had put it, "run rings around the council" was part of the appeal, "that's what I love about this job; I am forced to constantly come up with new ways to do this the best way".

## The limits of selling the periphery: lifestyle living in Gisborne, VIC

In early 2022, the 18th season of *The Block* was facing a string of protracted difficulties in its new peri-urban setting. First running for two seasons between 2003 and 2004, before a revival in 2010 with a new season every single year since, the Australian reality TV show (New Zealand has its own localized spinoff with ten released seasons) follows four or five couples as they engage in competitive home renovation according to a particular set of themes or challenges, ending with an auction for each of the houses, where of course the goal for each of the competing couples is to sell their renovated house at the highest possible price. Successive seasons have focussed on apartments, villas, terraced rowhouses, and the conversion of commercial land, hotels, or office space into luxury housing, all within the urban confines of Sydney and Melbourne. In late 2021, for the first time, news reports from Nine, the company that produces the show, told fans that *The Block* will be seeking rural lifestyle blocks for renovation and sale:

> It's literally going to be the biggest Block ever, with the teams tackling a whopping 10 acres each in Victoria's Macedon Ranges. There will be tractors, buggies, excavators, rolling green hills and plenty of land, giving fans of the show something we've never seen before.
>
> (Lyall, 2021)

Over the same period of time as my fieldwork in Kumeu-Huapai, however, construction of these lifestyle block residences in Gisborne, some 50 km north of Melbourne, began to run into a "perfect storm" of barriers to their work. Materials became "unobtainable" or tripled in price (Brook, 2022), while the availability of flexible labour plummeted, leading the executive producer to bemoan "an Australia-wide shortage of tradies" and send out a "desperate plea" for any available construction workers to start work on the site immediately and work crunch hours for the five weeks remaining before the date of the auction and thus the end of production (Fowler, 2022). Flooding caused general delays to work and shipping and even the Russian invasion of Ukraine led to some forms of laminated timber imported from the offending nation to have tariffs added due to its status as "conflict timber". Materials that once could be ordered at 9 am and brought to the peri-urban build site by 11 am the same day now take upwards of a month (Brook, 2022), cost up to three times as much as before, and public appeals had to be made (over media sites owned by the same company that runs this production) for skilled workers to head to Gisborne and save the show.

This rather sudden and severe series of barriers to the smooth operation of the "massive machine" that can transform a rural land bank into a highly profitable lifestyle living and then an unbeatably profitable suburban development is particularly nerve-wracking for land developers and their financiers. As the lawyer in the previous chapter put it "nothing hurts a margin more than expenses you didn't budget for, delays and obstructions". The very first chapter of this book touched upon its impact on the new developments at Auckland's northwestern frontier – long backorder times for vital materials, smaller housebuilders finding themselves driven out of the market as large franchises clamp down on the scarce amounts of available raw material – but the picture for residential construction in Melbourne and much of urban Australia is even more fraught, for reasons that will be examined over this section. Individual construction companies, pushed to the margins of viability by rising property prices and profitability requirements even under the best of circumstances (Murphy, 2020, p. 1510), are beginning to fold in Melbourne's state of Victoria and across urban Australia (Sharples, 2022) leading to understandable fears from the sector of a generalized industry collapse. When one Melbourne-based operation declared insolvency in early July, former staff members told the media that in the period before liquidation, 60 per cent of their projects had been failing to turn a profit (Turner-Cohen, 2022). This failed company had specialized in boutique residential developments to landowner specifications in affluent areas of suburban Melbourne, the same form of development which defines many of the new suburbs of Kumeu-Huapai. To examine commonalities and useful points of difference between the business of peri-urban development today in Auckland and the same work being undertaken in Melbourne, I will look at suburban and lifestyle development around Gisborne, northwest of the Melbourne City centre (see Figure 5.4). While similarities in financialized development between the cases are glaring, important differences in scale, in forms of investment, and in

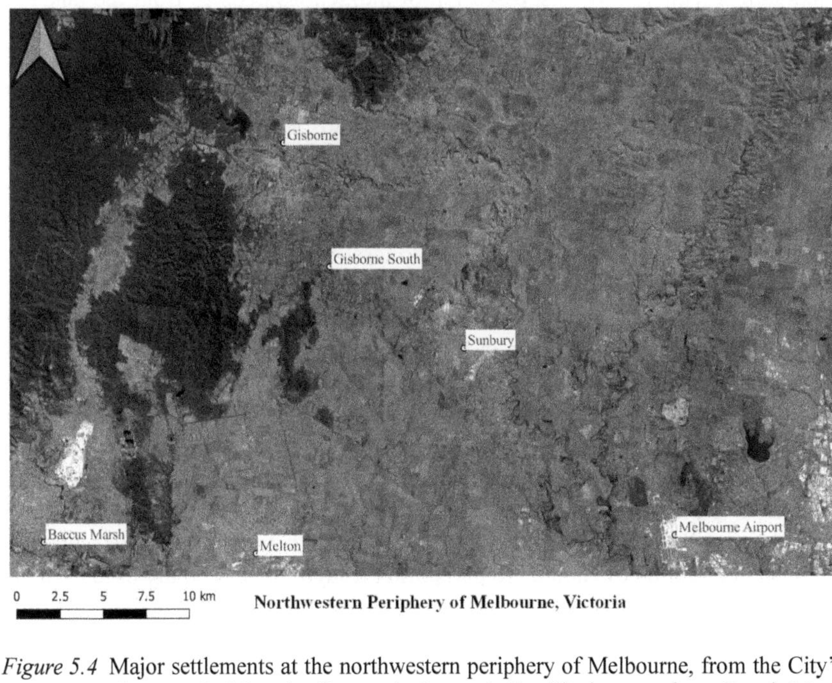


*Figure 5.4* Major settlements at the northwestern periphery of Melbourne, from the City's airport to the Macedon Ranges in the north. Satellite imagery from Google Maps (Google LLC, 2022).

the relationship between development and the changing climate make this slice of peri-urban Melbourne useful in understanding how Kumeu-Huapai relates to other formerly rural towns at the edge of growing cities across the world.

Auckland is the largest urban area in the entirety of Aotearoa by far, containing more than one-and-a-half times as many people as the entirety of the South Island (Stats NZ, 2022). Melbourne, including its own exurban frontier, has a similar population to all of New Zealand at some five million residents. Auckland has been projected to grow rapidly (not just in surface area), reaching 2.4 million people by the middle of the 21st century, some 40 per cent of the nation's population (Auckland Council, 2018, p. 9). By that same time, projections estimate that Melbourne will have grown to contain eight million people (Victoria State Government, 2017, p. 7). Auckland and Melbourne both claim the dubious honour of being among the most unaffordable cities to purchase a home in on the planet (Urban Reform Institute and the Frontier Centre for Public Policy, 2022), though Melbourne sits a few places higher on the list – closer to Hong Kong and London – due to its more notable popularity as a site of international investment in real estate, even if (as in Auckland) the "foreign" nature of real estate investment is often overstated for political purposes (Rogers et al., 2015). Melbourne's own peri-urban frontier is thus home to a much larger example of the very same phenomenon presented in the first half of this book.

Naarm/Melbourne carries a similar colonial history of land use as Tamaki Makaurau/Auckland. The dispossession of indigenous peoples from the lands at the new colonial frontier in the 19th century was initially carried out in the name of speculative ventures rather than for production. This was especially so during the first settlement rush of the early 1840s (Edmonds, 2010, p. 134) and the "land boom" in colonial Melbourne over the 1880s and 1890s. In the latter case, land parcels saw, on average, annual appreciation rates as high as 78 per cent in some years (Silberberg, 1975, p. 214). However, as seen in the lives of two Sydney-based men named William Deloitte in Chapter 2, processes of colonial appropriation saw Auckland as more peripheral than those in the growing cities of Australia. The purchases of blocks in Taupaki and arrangement of mortgages for Waikoukou were sent from the desk of the younger Deloitte from the auction company his father had founded in Sydney. In the years when the population of colonial Auckland numbered a few thousand, and the Kumeu River Valley was still used entirely as a place to harvest Kauri wood and to sell subdivisions to new arrivals to colonies, Melbourne banks were already having massive regular financial crises of speculation on land and commercial interests in the young state of Victoria (Hall, 1968). For these reasons, peri-urban development in Melbourne can help us understand how Auckland's own situation is defined by the scale and by the concentration of capital. To paraphrase a saying occasionally used to denigrate New Zealand's biggest city in comparison with Sydney, Auckland is Melbourne for beginners. Should the trend towards urbanization and financialization continue, what happens at Melbourne's periphery now can teach us about Auckland's future.

As well as being comparable as colonial cities deemed favourable for land speculation and other financialized forms of urban development, the two urban centres share a trait found in many growing cities: their settlement sits in the middle of a regional foodbowl. Half of the vegetables eaten in the state of Victoria come from Melbourne's foodbowl, and while this productive capacity currently corresponds with more than 80 per cent of Melbourne's needs, the continued urbanization of the city's periphery will reduce this to less than 20 per cent by 2050 (Sheridan et al., 2015, p. 3). As referenced near the end of the previous chapter, the threat of formerly productive places like Melbourne becoming net importers of food in the foreseeable future (Turner et al., 2018) has led to a series of policy proposals – still entirely hypothetical – that consider top-down and inflexible restrictions on the fragmentation of productive land in order to quash speculation and finance-driven forms of suburban development (Buxton & Butt, 2018; Carey et al., 2018). Under the current arrangement between state and market interests, greenfields development dwarfs intensification in the inner suburbs (Buxton et al., 2016, p. 85), and the economic underestimation of the broad range of services offered by productive peri-urban land tends to, as in the case of Auckland, favour the release of land for development (Buxton & Butt, 2018, p. 108). Also like Auckland, this dynamic is self-perpetuating, as the flexibility of future zoning provisions makes land speculation more viable and investment in peri-urban food production less so, and so on (Carey et al., 2018). It is in this familiar context that we turn to Gisborne, a fast-growing formerly rural town now at the frontier of "lifestyle block paradise" for Melbourne commuters.

Speaking to a planner working for the Macedon Ranges Shire Council, the regional body with jurisdiction over Gisborne, Victoria, I was made aware that there are "significantly more rural living zones around Gisborne" compared to other green wedges beyond the Melbourne metropole. Over the videocall, they pointed to a large swathe of rural land surrounding the Gisborne and reaching down towards rapidly urbanizing Sunbury (see Figure 5.5), explaining that these had been transformed from rural production to lifestyle living in almost a single motion around the turn of the millennium, irreversibly transforming this stretch of land from suburban Melbourne into the foothills of the Macedon ranges into only being suitable for subdividing into lifestyle blocks.

While adverts for particularly expensive Gisborne rural lifestyle properties, again published by the same media company that runs *The Block*, promise that the TV show will in the future "put Gisborne South on the map for property aficionados" (Power, 2022), the very same article will note that the property in question is already on the market for a sum $500,000 AUD (Australian Dollar) higher than that listed only a year ago and that entire Macedon Ranges Shire has seen "rocketing 12-month house price growth". A suburban house in Gisborne that sold a decade ago for $200,000 AUD is now valued at over a million (Property Value Australia, 2022a), and a 3.5 Ha lifestyle property that went to auction for $900,000 AUD eight years past is now considered to be worth anywhere up to four million (Property Value Australia, 2022b). Just as in Auckland, the popularity of lifestyle living for Melbourne-based professionals has only increased, and Gisborne's glut

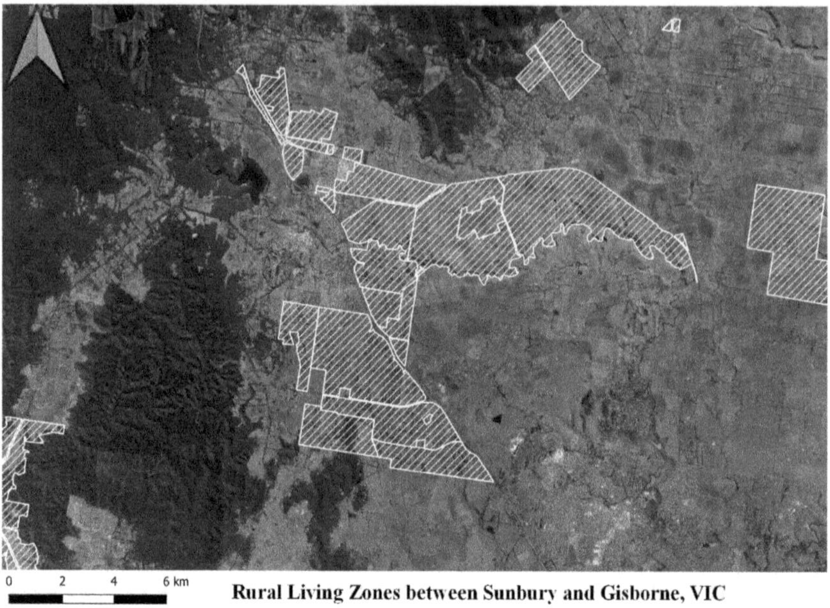


*Figure 5.5* Rural Living Zones northwest of Melbourne, between Gisborne and Sunbury (Victoria State Government, 2022). Satellite imagery from Google Maps (Google LLC, 2022).

of lifestyle-zoned land makes it a peri-urban destination of choice. In fact, one reporter covering *The Block* tells readers:

> The idea to take *The Block* on a tree change, like so many Aussies have done in the past few years, was inspired by executive producer Julian Cress' own move to Gisborne South with his family during the COVID-19 lockdowns.
> (Aubusson, 2022)

In much the same way as the speculative interest in Kumeu-Huapai and other frontiers of Auckland helped in turn spur further financialized development, the promise to "put Gisborne on the map" is built on the financial momentum already being generated by Melbourne residents seeking a rural aesthetic.

A promotional video for the new season of *The Block* begins with a large new four-wheel-drive gliding along a single road among pristine green hills while a voice-over narrator announces, "this is the new Aussie dream, a gigantic slice of pure heaven". Once the car pulls up at the building site and the long-time host proclaims "welcome to Block Country", the host and narrator both explain that the conceit of the season will be bringing five couples of "City Slickers" out to the countryside, along with a joke about a woman in high heels stepping in cow manure and promises to the audience that on this "prime Aussie heartland, as far as the eye can see", tennis courts, wineries, and decorative lakes will be constructed.

As pointed out at the opening of this section, however, such an ambitious (and certainly very profitable) transformation of the landscape to reorient it towards the desires of the city-based professionals is very contingent on the constant supply of cheap labour and materials drawn in from across the planet. Attempts by the Australian federal and state governments to spur an economic recovery from the COVID-19 pandemic by encouraging property development (Better Regulation Victoria, 2021, p. 12) as well as the phenomenon of real estate being the final sanctuary of profitable investment, discussed earlier in this book, have served to create an unprecedented demand in the past two years for both construction materials and the skilled and mobile labour needed to actually build houses in peri-urban Melbourne and rural Victoria (p. 3).

Along with this dramatic increase in demand has come disruptions to international supply. The first year of the pandemic saw the same waves of disruptions felt by housebuilders in Auckland's northwest in Chapter 1 of this book, with some 60 per cent of construction materials (by dollar value) coming directly from China, at least according to Australia's construction industry lobby at the outset of the pandemic (Australian Construction Industry Forum, 2020). Similarly, barriers to movement created by the pandemic reduced Australia's access to cheap labour. By the beginning of 2022, vacancies in the construction industry had increased by 80 per cent compared to 2019, leading to an estimated 100,000 unfulfilled roles by 2023 (Arcadis, 2022, p. 15). And of course, right on time for the showrunners of *The Block*, the fact that half of Australia's Laminated Veneer Lumber is imported from Russia became very relevant for the residential construction supply chain. Once Russian armoured vehicles crossed the border into Donbas, their flanks spraypainted with haphazard Zs to differentiate them from identical military equipment owned by Ukrainian forces, importers in Australia began to assume that disruption of supply due to war could be almost total, even without considering the tariffs applied to the "conflict timber" which residential construction industry relied on (Sadauskas, 2022). The very delicate circumstances that enabled the existence of the residential construction boom in the Australian state of Victoria have run up against painful, unpredictable, and unavoidable limits.

"You work with the decisions of the past" the planner remarked, acknowledging that all subsequent decisions over land use in the area today must be made in context with rulings over zoning that had been made two decades ago. Some towns in the Macedon Rangers Shire, they told me, were mandated to have zero new construction for accommodating population growth. The reason for the Council to impose such inflexible constraints even when the state government requires that local authorities create new housing capacity at the periphery is that this rapid expansion of suburban life has run up against another limit to expansion that cannot be smoothed over. The release of large tracts of rural land for residential living at the frontier of Melbourne's urban economy over the new millennium bought the new households of Melbourne professionals directly into the path of Victoria's wildfire season. In 2009 (and more recently in the massive fires of late 2019), as Buxton, Goodman and Moloney put it, "The high rate of peri-urban population increase and the addition of thousands of detached dwellings to small rural lots, many close to vegetation, placed many people within a 'colossal fire flume'" (Buxton et al., 2016, p. 93). As the peri-urban Shire with the most Rural Lifestyle Zoned land by far (more than 30 per cent by area, while no other Shire has more than 10 per cent [Buxton et al., 2011, p. 7]), the Macedon Ranges – and thus Gisborne – is particularly vulnerable to an overextension into territory that simply cannot accommodate year-round human habitation. The only solutions come from the kind of top-down anticipatory planning regimes that are "the antithesis of business-as-usual practices typically characterised by reactive, market-oriented policies enabling incremental, ad-hoc development" (Buxton et al., 2011, p. 2). Decades of wildly profitable fragmentation of Melbourne's peri-urban landscape has finally brought it up against a genuinely deadly limit to expansion.

Closer to the City Proper, in exurbs like Sunbury that are becoming incorporated into Melbourne's urban mass, the limits on profitable development would be much more familiar to my interlocutors across Kumeu-Huapai. "Some of these areas, the shops aren't coming in, the schools aren't either" the Macedon Ranges planner explained. One article from *The Guardian*, an investigation of family life in the new urban frontier of Melbourne, presents the testimony of a couple who, like the French professionals in Bourdieu's study of the housing market in Chapter 1 of this book, were sold suburban life and found they had been misled. "Since the couple moved in, plans for a local primary school have been derailed, and plots of land still lie empty while residents wait for shops, services and infrastructure" (Visontay, 2021). The title of the article, "A broken dream: outer Melbourne has affordable houses but no train or school" is as stark a description of the future of the formerly rural towns at Auckland's frontier today as has been seen in this book. The provision of these new residential developments on the promise of services that never materialized is both a perfect case of Bourdieu's dramatically named "foundations of petit-bourgeois suffering" and is a useful representation of the future of Auckland's new suburbs should the urban periphery remain chiefly valued as one of the few remaining sites of safely profitable investment. The article from *The Guardian* quotes planners who see such an obviously profit-orientated approach to development as a result of their failure to keep capital in check, "inadequate planning laws have allowed developers to dictate the lifestyles of a generation of future Melburnians" (Visontay, 2021). In the same way that Carey and Sheridan have been cited earlier in this section as demanding long-term planning restrictions to cut down on land fragmentation, the failure of Melbourne's new suburbs to provide the services needed for human life is leading officials to reorient state interests away from being in concert with those of developers.

On a larger scale and at a greater concentration of capital than Auckland's own speculative acceleration of residential sprawl, the barriers and limitations created by the massively profitable urbanization of Melbourne's periphery provide an insight into the challenges that places like Kumeu-Huapai will face in the near future, as well as corroborating the details of the phenomenon that drives land fragmentation and sprawl in Auckland. The two notable barriers to further profitable accumulation in rural lifestyle areas such as Gisborne – the instability of the supply of cheap labour and materials from international sources, and the dangers to human life posed by expanding into areas where safe continuous habitation cannot be guaranteed – will only become more relevant at Auckland's northwestern frontier (and many other growing cities) in coming years, and the unfulfilled promises of services and infrastructure for new suburbs in Outer Melbourne are already a concern for developments within driving distance of Auckland City. Having used the outskirts of Melbourne as an example of consequences and complications of "selling" the periphery, the following section will look at how extremely detrimental conditions such as inequality and pollution can be simply the price of making a fortune in Fresno, California.

## A license plate reading "REZONE": the relationships between Fresno's residents and its dominant industries

"West Fresno has been asking for years, if not decades, for single-family, market-rate housing". Fresno council member Miguel Arias is announcing that the city and a dairy farm have come to an agreement to convert some of the latter party's land into 600 single-family suburban homes and a half-hectare public park. While "West Fresno" in the abstract is seeing a lot of demand for new maximally profitable forms of housing, the people who live and work there have not necessarily desired for decades, above all else, to see new suburban market rate housing be built. Cassandra Garibay, writing on the Council's announcement in the Fresno Bee, pointed out: "48 per cent of people in the census tract where the housing development will be located are living below twice the federal poverty level". Garibay noted that Ivanka Saunders, a local policy advocate, argued against the move by saying "The city should hold developers accountable to closing the gap of homeownership by uplifting and coordinating the available programs … the original residents of west Fresno deserve the opportunity to be homeowners" (Saunders, 2021). For the final comparative case of this chapter, I will examine the relationship between the individuals who make their fortunes in the business of property development (as well as few other major local moneymakers) and the people who actually live in the parts of Fresno, California, which offers the most profitable development opportunities for developers. The transformation of the peri-urban landscape into one most suited for profitable suburban development necessarily prioritizes the desires of the developer over those of the people who make their lives at the edge of the fast-growing city and do not happen to be major landowners or financiers.

When Hurricane Katrina made landfall in New Orleans at the end of August in 2005, it left 80 per cent of the city and large swathes of outlying parishes flooded for weeks on end, and necessary services such as clean water and electricity would not be restored to some areas for years. Yet, when the Brookings Institute published a post-Katrina study of concentrated poverty in major US cities as part of an effort to examine the marked racial disparity in the impact of the storm, New Orleans was not the city with the highest deprivation and poverty. The devastated City was certainly high on the study's metric of "concentrated poverty", as in how many desperately poor people live within a single neighbourhood census area, but it was second to Fresno, one of the most productive agricultural hubs in the entire United

States (Benner & Pastor, 2015, p. 139). Located in the San Joaquin Valley, Fresno, California, is roughly equidistant from San Francisco and Los Angeles, about 200 miles by car to either one of these major urban centres. The three richest industries in Fresno – Industrial-Scale Agriculture, Suburban Property Development, and Oil Extraction – above all require access to cheap labour and the removal of environmental regulations that would restrict profitable operations (Benner & Pastor, 2015, p. 143; Cannedy, 2020, p. 7). Under the demands of such interests, the poverty rate doubled since 1980 and 2010 (Benner & Pastor, 2015, p. 140), and indications of significant social and environmental injustice such as localized air pollution and food insecurity are among the highest in the United States (Fairbanks, 2021, p. 14; Padula et al., 2018). This burden is not evenly spread, of course. Should anyone look at the broad expanse of Fresno's sprawl with metrics for poverty, air pollution, low birth weights, asthma, hazardous waste dumping, and many others, there is a

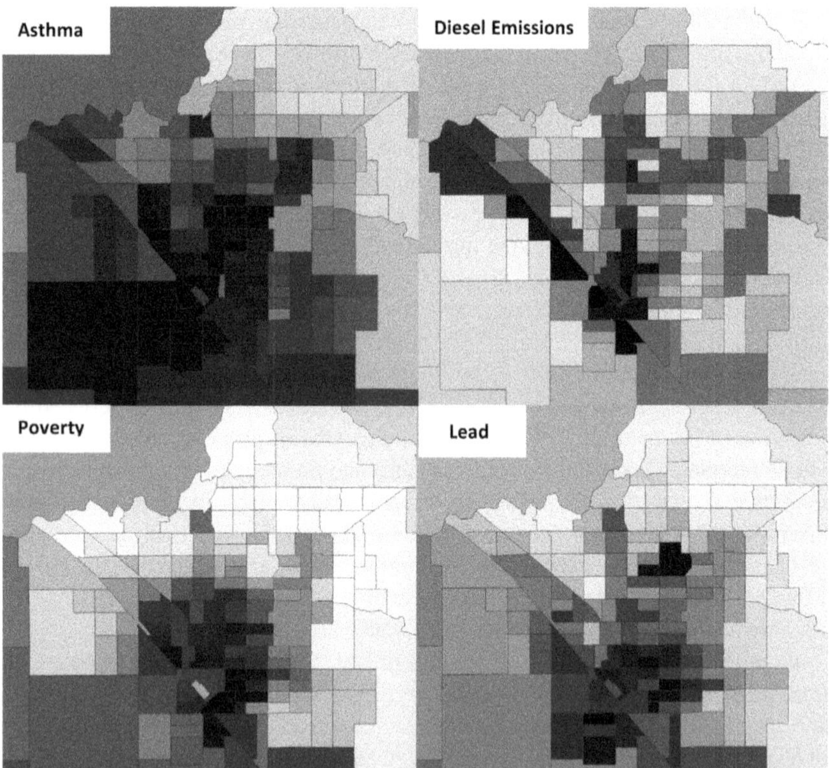


*Figure 5.6* Examples of the NE-SW segregation of Fresno districts as seen from various CalEnviroScreen metrics (California Office of Environmental Health Hazard Assessment, 2021). Scales for every one of these phenomena are different, so I have set a general scale from White to Black, where Black is the highest incident of the social/environmental ill in question.

clear divide between Fresno's northeastern periphery and its western and south-western reaches (see Figure 5.6). Fresno's political economy is built on this particularly stark segregation on lines of class, race, and physical space, between those who make money from the three industries and those who live and work under the social and environmental costs of their profitability.

Agricultural yields are declining, even if Fresno remains the biggest producer in the state (California Department of Food and Agriculture, 2020, p. 21), and while Fresno has 2,000 oil wells, the third highest amount for any county in California, Neighbouring Kern County dominates the business with 42,000 active wells (Benner & Pastor, 2015, p. 142). If there is a growth industry in Fresno, it's the remaining member of the triumvirate: real estate. Compared to the coastal urban centres of California, the relative affordability of land and housing in suburban Fresno led to a rush of speculative purchases in the property bubble that precipitated the 2007–8 Global Financial Crisis. When the dust from said crisis had settled, Fresno stood out for the extreme volatility of financial valuations for its (mostly single-family suburban) housing stock. After increasing by nearly 150 per cent between the new millennium and the end of 2006, Fresno's house price index fell to just above pre-bubble levels by 2010 (Federal Housing Finance Agency, 2021). One Kumeu-Huapai real estate agent had once explained the relative security of their supposedly quite volatile industry by telling me "We do well at the top and at the bottom of the market". So, it was in Fresno, particularly western and south-western Fresno, from 2007 onwards. Foreclosure sales and "short sales" – where a homeowner is convinced to sell their house at a discount and be left with the remaining debt rather than be foreclosed on – were unusually severe in Fresno compared to the rest of the United States. Short sales were made at discounts 13 per cent bigger than the national average, while foreclosed homes were sold on the post-crash housing market a full 20 per cent more discounted than seen across the rest of the country (Aroul & Hansz, 2014). The poverty and pollution placed on western Fresno by the region's dominant industries made it a promising candidate for speculation on very affordable land and housing in a time where property values had been skyrocketing nationwide. When the bubble burst, investors could take the market coming the other way, picking up houses in suburban Fresno from its especially economically vulnerable residents at unbeatable prices. Here, far clearer than in Kumeu-Huapai or the Macedon Ranges, we can see the "creative destruction" (Harvey, 2007) necessary to the process of the enrichment of property owners and financial agents.

The property bubble of the early 2000s was by no means Fresno's first prominent billing as a site of deeply exploitative property speculation practices. An extremely close relationship between city officials and developers, the intimacy of which makes the links between New Zealand's National Party and the country's developers seem absolutely circumstantial by comparison, has defined the fragmentation and urbanization of Fresno's agricultural periphery for decades. Strategies used to allow rural land, which is actually protected under various state legislature measures intended to encourage peri-urban production (County of Fresno, 2022), to be broken up and developed for maximized private profit, range from the more casual connections described elsewhere in this book to open and direct bribery. These practices were (and still are) so blatant that in 1994, an FBI sting was launched specifically to crack down on what amounted to a criminal conspiracy operating across the territorial field of Fresno's urbanizing periphery. The undertaking was christened Operation: REZONE after the custom license plate of the vehicle owned by a somewhat overconfident local real estate agent with ties to the local government (Benner & Pastor, 2015, p. 144; Zuk, 2013, p. 52). Today, the city council promises to "resolve" the dispute between poor residents and wealthy landowners by removing all restrictions on land use in the area (Ugwu-Oju, 2021), to block attempts by the state legislature to demand less profitable forms of urban development (Garibay, 2021) and to forestall the introduction of an alternative plan for the city supported by disadvantaged communities and small horticulture operations (Benner & Pastor, 2015, p. 149).

That local politics and the desires of developers and significant polluting interests would be so closely aligned becomes far less confusing when one looks at an award-winning undergraduate thesis from Cade Cannedy (2020) of Stanford University, who had simply sought to compare the career backgrounds of air pollution regulators in Fresno County. They found that more than half of the people who sat on positions in a local government dedicated to regulating pollution (from, again, major local industries such as oil extraction and industrial-scale farming) had direct and ongoing conflicts of interest with the polluting groups they were charged with regulating (p. 58). In the business of property development, the arrangement is comparable. A real estate agent in Kumeu-Huapai may have rather casually let me know that they had made some lucrative deals a decade back by simultaneously overseeing the rezoning of the town during the reorganization of Auckland's local governing structures, but the scale of such corruption is dwarfed by the almost total integration of developer lobbying groups such as Building Industry Association of Fresno into decision-making processes across the county. These groups block attempts to reduce sprawl or fragmentation (Benner & Pastor, 2015, p. 142), something the Council considers "the restriction of free enterprise" (Zuk, 2013, p. 38). Development interests and other social and environmentally destructive industries have aligned Fresno's local policy to their ends long before the first property bubble of the 21st century, and this close relationship between ruling interests and policy in the growing city will persevere well into the future.

The unmatched profitability of development at Fresno's periphery is made possible not only by a dramatic segregation of the population into profiteers and the victims of long-spanning social and environmental crime but also by the phenomenon that has occupied so much of this book – the slow destruction of urban foodbowls. Yet, the loss of California's productive horticulture – with all the globe-spanning ramifications that would entail – is being driven not just by the unending demand for land to convert into suburban housing. The destabilizing climate of the entire region and the responses of large farming operations to the ever-more severe droughts being wrought by this change both act as a powerful stressor on the viability of peri-urban horticulture and thus serve to accelerate the loss of productive land to speculative purposes. Just as in formerly productive areas of Auckland and Melbourne, residents and officials in Fresno are concerned by the loss of productive land to suburban sprawl – “Could we actually look up in 30 years and wonder where all our farmland – and food – went?” (Toste, 2021). Attempts to predict the spread of lifestyle blocks and suburban developments present stark warnings of the loss of hundreds of thousands of hectares of prime growing land by the middle of the century (California Department of Food and Agriculture, 2015, p. 5). At the same time, droughts have forced agricultural practices across the state to change drastically, and the dominant responses have only made production more vulnerable (Christian-Smith et al., 2015). As rainfall failed to appear, farmers pumped more groundwater. As hydroelectric dams began to dry up and stop generating power (p. 495), operations switched to fossil fuel-powered generators. Many simply allowed land to go fallow to reduce water consumption. The drought of 2007–9 alone drained more groundwater than the entire capacity of Lake Mead (p. 498), increased carbon emissions state-wide by 10 per cent (p. 497) and led to bail out packages being paid by the state to 200 farming operations in Fresno County (p. 499). The result of these droughts is a cascade effect (Tortajada et al., 2017, p. 100) – power generation and agriculture are jeopardized in the short term, then the production of fruit and vegetable is abandoned in favour of crops (or other land uses) that need less water, farms find it hard to stay afloat, the stress placed on groundwater by drought response, in turn, destroys the viability of horticulture operations, and so on. These droughts also bring more frequent wildfires, which, like those seen at Melbourne’s urban periphery, are a uniquely immediate threat to suburban life and will be discussed in detail in Chapter 8. At Fresno’s urban edge, the complete co-option of governance interests to the FIRE economy collides with the literal firestorm of an unravelling biosphere.

## Total urbanization and global suburbs: the expansion of the financialized periphery

When rural production in Kumeu-Huapai collapses in the face of market-oriented reforms and land speculation, when dominant forms of suburban development make landowners and developers rich while failing to provide basic infrastructure and services needed for working life, and when the city-based professionals who move into these new suburbs find their home was built in the path of worsening natural disasters, people living and working at the frontier of expanding cities across the planet might find the situation familiar. Just as the same developers and landowners who are transforming Kumeu-Huapai also look further afield in Aotearoa for speculative opportunities, individual capitalists and professionals in Melbourne and Los Angeles seek out rural land and regional towns that offer real estate at far more affordable rates than can be found at the commanding heights of the urban economy, and in doing so, these agents bring the dynamics of the financialized urban periphery out well beyond the reach of any harried professional’s daily commute. Not only is the case presented in Part One of this book a common occurrence, but it is becoming applicable to the lives of more people every year, as the extension of the financialized periphery by development agents renders entire states the territory of the city and brings suburban ambitions to every corner of the planet (Davis & Monk, 2011).

Kumeu-Huapai, a town of a thousand, then suddenly 5,000 people, only 20 km from Auckland City centre, with unparalleled property value rises between 2014 and 2019, is a particularly dramatic and ongoing case study of how materially valuable but less profitable practices such as small-scale peri-urban horticulture can be swept aside in favour of abstract financial accumulation. Yet, other examples of this phenomenon, as seen in this chapter, can provide clues as to how Kumeu-Huapai and other formerly rural towns at Auckland's urban frontier will grow and change, and what challenges their residents will face in the future. For this reason, three notable areas of contrast across the international case studies presented here will form the basis of each of the three remaining chapters of this book. First, the following chapter will focus specifically on the class (and indeed caste) dynamics that define profit-making in Kumeu-Huapai and in other new suburbs serving the urban centre, with the particularly blatant segregation of cities like Fresno in mind. After that, the relationship these growing cities have to their own food security will be discussed, as settlements once founded in areas valuable for food production seem to become ever hungrier for any nearby land to convert into the maximally profitable land use of suburban residential developments. Finally, the last main chapter of this book will confront the future of peri-urban finance-led development, in a world where land suitable for either growing food or for building houses on becomes increasingly scarce as climate change and urbanization collide.

## Bibliography

Andrew, M. (2021). *Ten Places Where You Can Still Buy a House for Under $300k*. Retrieved from https://thespinoff.co.nz/business/17-06-2021/ten-places-where-you-can-still-buy-a-house-for-under-300k

Arcadis. (2022). *International Construction Costs 2022*. Retrieved July 13 from https://www.arcadis.com/en/knowledge-hub/perspectives/global/international-construction-costs

Aroul, R. R., & Hansz, J. A. (2014). The valuation impact on distressed residential transactions: Anatomy of a housing price bubble. *The Journal of Real Estate Finance and Economics*, *49*(2), 277–302.

Aubusson, L. (2022). *First Look at New Season of The Block: Tree Change in Victoria's Macedon Ranges*. Retrieved July 13 from https://www.kidspot.com.au/lifestyle/entertainment/first-look-at-new-season-of-the-block-tree-change-in-victorias-macedon-ranges/news-story/08a3f680c56ed41f29edc1d7ebeea27e

Auckland Council. (2018). *Auckland Plan 2050 Evidence Report: Demographic Trends for Auckland: Data Sources and Findings June 2018*. Auckland Council. Retrieved from https://www.aucklandcouncil.govt.nz/plans-projects-policies-reports-bylaws/our-plans-strategies/auckland-plan/about-the-auckland-plan/Evidence%20reports%20documents/evidence-report-demographics.pdf

Australian Construction Industry Forum. (2020). *ACIF Urges Industry and Government Collaboration to Address the Threat of the Coronavirus to the Construction Industry*. Retrieved July 13 from https://www.acif.com.au/acif-news/acif-urges-industry-and-government-collaboration-to-address-the-threat-of-the-coronavirus-to-the-construction-industry

Benner, C., & Pastor, M. (2015). *Equity, growth, and community: What the nation can learn from America's metro areas*. Oakland: University of California Press.

Better Regulation Victoria. (2021). *Addressing Supply Chain Challenges-Review into Issues Facing Victoria's Building and Construction Industries*. Retrieved from https://www.vic.

gov.au/sites/default/files/2022-04/Addressing%20Supply%20Chain%20Challenges%20-%20Final%20Report%20-%20website.pdf

Brook, S. (2022). *The Block Rocked by Home Building Crisis*. Retrieved from https://www.theage.com.au/politics/victoria/the-block-rocked-by-home-building-crisis-20220526-p5aot4.html

Buxton, M., & Butt, A. (2018). *The future of the fringe: The crisis in peri-urban planning*. Melbourne: CSIRO Publishing.

Buxton, M., Goodman, R., & Moloney, S. (2016). *Planning Melbourne: Lessons for a sustainable city*. Melbourne: CSIRO Publishing.

Buxton, M., Haynes, R., Mercer, D., & Butt, A. (2011). Vulnerability to bushfire risk at Melbourne's Urban Fringe: The failure of regulatory land use planning. *Geographical Research*, *49*(2), 1–12.

California Department of Food and Agriculture. (2015). *Benefits of Farmland Conservation in California*. Retrieved from https://www.cdfa.ca.gov%2Foefi%2Fclimate%2Fdocs%2FFarmland-Conservation-in-California.pdf

California Department of Food and Agriculture. (2020). *California* Agricultural *Statistics Review 2019–2020*.

California Office of Environmental Health Hazard Assessment. (2021). *CalEnviroScreen 4.0*. Retrieved from https://oehha.ca.gov/calenviroscreen/report/calenviroscreen-40

Cannedy, C. (2020). *Why Your Air Will Not Get Better: Path Dependence and Capture in Air Quality Regulation*. Stanford. Retrieved from https://west.stanford.edu/works/why-your-air-will-not-get-better-path-dependence-and-capture-air-quality-regulation

Carey, R., Sheridan, J., & Larsen, K. (2018). *Food for thought: Challenges and opportunities for farming in Melbourne's foodbowl*. The University of Melbourne. Retrieved August 8 2021 from https://science.unimelb.edu.au/foodprint-melbourne/publications/report-challenges-and-opportunities-for-farming-in-melbournes-foodbowl

Christian-Smith, J., Levy, M. C., & Gleick, P. H. (2015). Maladaptation to drought: A case report from California, USA. *Sustainability Science, 10*, 491–501.

Corlett, E. (2018). *Pukekohe Grower: 'We're Really the Last Frontier'. Radio New Zealand*. Retrieved August 15 from https://www.rnz.co.nz/news/national/364134/pukekohe-grower-we-re-really-the-last-frontier

County of Fresno. (2022). *California Land Conservation Act or Williamson Act*. Retrieved July 18 from https://www.co.fresno.ca.us/departments/assessor/real-property-assessments/williamson-act

Davis, M., & Monk, D. B. (2011). Evil *paradises: Dreamworlds of neoliberalism*. New York: The New Press.

Deloitte. (2019). *New Zealand's Food Story: The Pukekohe Hub*. Retrieved from https://www2.deloitte.com/nz/en/pages/primary/articles/case-study-pukekohe-hub.html

Desmarais, F. (2022). *Rotorua Reserves for Housing: Council Urged to Rethink Sale of Community Assets*. Retrieved from https://www.rnz.co.nz/news/ldr/470184/rotorua-reserves-for-housing-council-urged-to-rethink-sale-of-community-assets

Dreaver, C. (2019). *Environment Minister: 'We Actually Prefer Potatoes'. Radio New Zealand*. Retrieved August 14 from https://www.rnz.co.nz/news/political/396668/environment-minister-we-actually-prefer-potatoes

Edmonds, P. (2010). The intimate, urbanising frontier: Native camps and settler colonialism's violent array of spaces around early Melbourne. In T. B. Mar & P. Edmonds (Eds.), *Making settler colonial space* (pp. 129–154). London: Palgrave Macmillan.

Fairbanks, D. H. (2021). Geography of hunger: Demographic and socioeconomic determinates of large-scale severe hunger rates in Fresno, California. *Journal of Poverty*, *25*(1), 1–22.

Federal Housing Finance Agency. (2021). *FHFA HPI County Map.* Retrieved from https://www.fhfa.gov/DataTools/Tools/Pages/HPI-County-Map.aspx

Forbes, S. (2022). *Auckland Council Set to Appeal Against Go-Ahead for New Town 'the Size of Napier' – Stuff NZ.* Retrieved from https://www.stuff.co.nz/auckland/300605951/auckland-council-set-to-appeal-against-goahead-for-new-town-the-size-of-napier

Fowler, B. (2022). *The Block Issues 'Desperate' Plea for Tradies Amid New Season Crisis.* Retrieved from https://www.news.com.au/entertainment/tv/reality-tv/the-block-issues-desperate-plea-for-tradies-amid-new-season-crisis/news-story/177f1e0589633669eb5f9962ab588034

Garibay, C. (2021). *Why Old Town Clovis, CA Won't Be Getting New Apartments-The Fresno Bee.* Retrieved July 18 from https://www.fresnobee.com/fresnoland/article255749376.html

Google LLC. (2022). *Google Satellite via QGIS.* Retrieved from http://mt0.google.com/vt/lyrs=s&hl=en&x={x}&y={y}&z={z}. Source: http://mt0.google.com/vt/lyrs%3Ds%26hl%23Den%26x%23D%27Bx%27D%26y%23D%27By%27D%26z%23D%27Bz%27D&zmax=18&zmin=20. Provider: wms.

Hall, A. R. (1968). *The stock exchange of Melbourne and the Victorian economy 1852–1900.* Canberra: Australian National University Press.

Harvey, D. (2007). Neoliberalism as creative destruction. *The Annals of the American Academy of Political and Social Science, 610*, 22–44.

Jeffares, M. (2022). *Experts Will Discuss the Value of the Taonga That Is the Heretaunga Plains.* Retrieved from https://www.nzherald.co.nz/hawkes-bay-today/news/experts-will-discuss-the-value-of-the-taonga-that-is-the-heretaunga-plains/D754MFOG4B3LZT3SLKTJ5FBJQY/

Kilgallon, S. (2021). *The Fraud Case That's Halted 'Monster' $1b Pōkeno Property Development.* Retrieved from https://www.stuff.co.nz/national/crime/123979299/the-fraud-case-thats-halted-monster-1b-pkeno-property-development

Kuprienko, D. (2022). *Homes in Whangarei Heads Snapped Up Sight Unseen.* Retrieved from https://www.oneroof.co.nz/news/40796

Lyall, A. (2021). *The Block is Returning in 2022—Here's What We Know So Far.* Retrieved from https://9now.nine.com.au/the-block/2022-season-block-confirmed-location-details-season-18-explainer/d120453c-1a67-460f-a839-cc0b8465e669

Milldale. (2022). *Masterplan.* Retrieved July 19 from https://www.milldale.co.nz/masterplan

Milldale Infrastructure LP. (2020). *Milldale Infrastructure Payments.* Retrieved June 30 from https://infrastructurepayments.co.nz/

Murphy, L. (2020). Performing calculative practices: Residual valuation, the residential development process and affordable housing. *Housing Studies, 35*(9), 1501–1517.

Naish, J. (2022). *Westport House Prices Continue to Rise Despite Flood Damage.* Retrieved from https://www.stuff.co.nz/national/128339226/westport-house-prices-continue-to-rise-despite-flood-damage

Olley, S. (2021). *Locals Devastated as House Prices Soar in Small NZ Town, Families Struggle to Find Rentals.* Retrieved from https://www.rnz.co.nz/news/national/448998/the-sad-price-of-new-zealand-s-soaring-housing-market

Padula, A. M., Huang, H., Baer, R. J., August, L. M., Jankowska, M. M., Jellife-Pawlowski, L. L., Sirota, M., & Woodruff, T. J. (2018). Environmental pollution and social factors as contributors to preterm birth in Fresno County. *Environmental Health, 17*(1), 1–21.

Power, E. (2022). *This Gisborne Glass House Gained $500,000 in 12 Months.* Retrieved July 13 from https://www.nine.com.au/property/news/this-gisborne-glass-house-gained-500000-in-12-months/e3c87cb9-2e24-4c76-b25c-b9915cc1fcd0

Property Value Australia. (2022a). *6 Coop Drive Gisborne-VIC 3437.* Retrieved July 13 from https://www.propertyvalue.com.au/property/6-coop-drive-gisborne-vic-3437/16465595

Property Value Australia. (2022b). *210 Mcgregor Road Gisborne-VIC 3437.* Retrieved July 13 from https://www.propertyvalue.com.au/property/210-mcgregor-road-gisborne-vic-3437/14541657

Radio New Zealand. (2019). *Building on Prime Growing Land-Nine to Noon.* Retrieved May 2 from https://www.rnz.co.nz/audio/player?audio_id=2018693232

Radio New Zealand. (2020). *New Zealand's Most Fertile Land Dug Up For Housing.* Retrieved from https://www.rnz.co.nz/national/programmes/first-up/audio/2018700986/new-zealand-s-most-fertile-land-dug-up-for-housing

Rogers, D., Lee, C. L., & Yan, D. (2015). The politics of foreign investment in Australian housing: Chinese investors, translocal sales agents and local resistance. *Housing Studies, 30*(5), 730–748.

Sadauskas, A. (2022). *Half of Australia's LVL Timber Comes from Russia – Who Knew, and What It Means for Construction.* Retrieved July 13 from https://thefifthestate.com.au/innovation/materials/half-of-australias-lvl-timber-comes-from-russia-who-knew-and-what-it-means-for-construction/

Saunders, I. (2021). *Hundreds of Homes Are Coming to West Fresno. Here's Why It's Been a Long Time in the Making.* Retrieved July 14 from https://www.fresnobee.com/fresnoland/article256300432.html

Sharples, S. (2022). *Two Victorian Building Companies Wulfrun Construction and Westernport Constructions Collapse.* Retrieved from https://www.news.com.au/finance/business/other-industries/two-victorian-building-companies-wulfrun-construction-and-westernport-constructions-collapse/news-story/018e0640fae760c4f229d5086eb08e7d

Sheridan, J., Larsen, K., & Carey, R. (2015). *Melbourne's Foodbowl: Now and at* Seven *Million.* Victorian Eco-Innovation Lab, The University of Melbourne.

Silberberg, R. (1975). Rates of return on Melbourne land investment, 1880–92. *Economic Record, 51*(2), 203–217.

Smith, C. (2022). *West Coast-Properties Snapped Up in Minutes and Buyers with Fistfuls of Cash.* Retrieved from https://www.oneroof.co.nz/news/west-coast-properties-snapped-up-in-minutes-and-buyers-with-fistfuls-of-cash-41075

Stats NZ. (2018). *Place Summaries-Kaikohe.* Retrieved from https://www.stats.govt.nz/tools/2018-census-place-summaries/kaikohe#work-income-and-unpaid-activities

Stats NZ. (2022). *NZ Stat.* Retrieved May 18 from https://nzdotstat.stats.govt.nz/wbos/index.aspx

Tortajada, C., Kastner, M. J., Buurman, J., & Biswas, A. K. (2017). The California drought: Coping responses and resilience building. *Environmental Science and Policy, 78*, 97–113.

Toste, J. (2021). *Farmland Disappears in Central Valley, More Planning Needed – The Fresno Bee.* Retrieved July 18 from https://www.fresnobee.com/opinion/readers-opinion/article249500410.html

Turner, G. M., Larsen, K. A., Candy, S., Ogilvy, S., Ananthapavan, J., Moodie, M., James, S. W., Friel, S., Ryan, C. J., & Lawrence, M. A. (2018). Squandering Australia's food security: The environmental and economic costs of our unhealthy diet and the policy Path We're On. *Journal of Cleaner Production, 195*, 1581–1599.

Turner-Cohen, A. (2022). *'Dream Gone': Customers Reeling as Another Victorian Builder Goes Bust.* Retrieved from https://www.news.com.au/finance/business/other-industries/dream-gone-customers-reeling-as-another-victorian-builder-goes-bust/news-story/652f8aba043f6fb1b247de406b87fce6

Ugwu-Oju, D. (2021). *Dyer Proposes Ordinance to Grant Fresno the Power to Rezone Land-Fresno Bee*. Retrieved July 18 from https://www.fresnobee.com/fresnoland/article256468276.html

Urban Reform Institute and the Frontier Centre for Public Policy. (2022). *Demographia International Housing Affordability 2022 Edition*. Retrieved March 24 from http://www.demographia.com/dhi.pdf

Victoria State Government. (2017). *Plan Melbourne 2017–2050: Metropolitan Planning Strategy*. Retrieved from www.planmelbourne.vic.gov.au%2F__data%2Fassets%2Fpdf_file%2F0007%2F377206%2FPlan_Melbourne_2017-2050_Strategy_.pdf

Victoria State Government. (2022). *VicPlan*. Retrieved July 13 from https://mapshare.vic.gov.au/vicplan/

Visontay, E. (2021). *A Broken Dream: Outer Melbourne Has Affordable Houses But No Train or School*. Retrieved July 14 from https://www.theguardian.com/australia-news/2021/nov/15/a-broken-dream-the-walkable-melbourne

Wetzstein, S. (2018). Comparative housing, urban crisis and political economy: An ethnographically based 'long view' from Auckland, Singapore and Berlin. *Housing Studies 34*(3), 1–26.

Zuk, M. Z. (2013). *Health equity in a new urbanist environment: Land use planning and community capacity building in Fresno, CA*. Berkeley: University of California.

# 6 Class, social allegiance, and race at the colonial urban periphery

*Figure 6.1* A real estate advert on Kumeu-Huapai's main road promising to convert urban prospective buyers to a rural life.

While the supply of urban labour is almost as important for the success of peri-urban horticulture as the urban demand for fresh produce, it was never as freely discussed by the former and current managers of small operations across the northwest. Walking into the back room of one of the extant horticulture operations to speak to the proprietor, I saw for the first time the half-dozen Pacific Islander women who actually do the fruit packing work, under the supervision of the owner I had been intending to meet. This owner told me they'd prefer to speak without the packers present, so when we did meet later, the warehouse empty, I asked about

DOI: 10.4324/9781003400325-9

these workers. Who were they? Where did they commute from every day? They told me "They are Pacific [Islanders] but they live here [in Auckland]. we use a contractor, he recruits ... most of them are cousins, family [*of the contractor*]". In other words, these women were Auckland locals, bought out to the periphery for the difficult and monotonous packing work often also handed off to young family members or seasonal workers, and their employment was based on a continuous relationship between the horticulturalist and a middleman, a relative who procures their labour.

This relationship between owners and labourers among small-scale horticulturalists is another contrast with larger and more isolated operations, who rely on specifically international sources of cheap labour to such a degree that they were at the forefront of demands to open New Zealand's borders for temporary workers during the first year of the COVID-19 pandemic. Yet, this does very little to change the obvious imbalance between the (white, either Anglo or Dalmatian) owners of horticulture operations in the northwest, and their labour force brought in from the Maori and Pasifika population of the city. When these owner-operators spoke of the significant amount of work they had personally put into family-owned horticulture plots from a young age, they were not lying about the long hours, the early mornings, and the sweat and dirt that had defined their teenage years on their parent's farm. What was instead notable was the absence of non-family employees, beyond occasional references to a relative placing advertisements that "gets some people ... lots of young people" for temporary work. This omission was equally notable among the managers of property development companies careful to draw a line between their business and the contracted labour scheduled in for specific on-site work. At the urban periphery of Auckland and other growing colonial urban centres, an important distinction is made by landowners in their narratives between the people who profit from the land, either from production or finance, and those who only labour to create these profits.

Decades after the liberalization of the economy rapidly accelerated class divisions and increased poverty rates (Kelsey, 2015, p. 87), New Zealand yet maintains a reputation as a remarkably equitable place to live. Even when the wealthiest 10 per cent own nearly 60 per cent of the nation's assets while the poorest half own some 2 per cent (Rashbrooke et al., 2021, p. 6), social life in this nation, as in Australia, is often described as being classless (France & Roberts, 2017, p. 15). There is in fact a significant contradiction in public opinion between the growing levels of inequality and class differentiation actually experienced by New Zealanders, and an unusually strong opposition to redistributive politics, at least at a broad electoral level (Nel, 2022, p. 24). At the core of this disjunction among the small-business owners, managers, and professionals who form the country's property-owning middle class is a genuinely widespread experience of inter-generational mobility in the second half of the 20th century (p. 34). For this group, who made up a significant portion of my interlocutors, early experiences of a relatively egalitarian life for European New Zealanders under the mid-century interregnum, with a state interested in maintaining high levels of employment and the subsidization of housing, allowed them to receive education, healthcare, and employment even without the help of wealth or family. When privatization began to take hold in New Zealand, this same cohort found opportunities in places where money was being made from liberalization, for example, those professionals mentioned in previous chapters who abandoned horticulture for a career in property. Today, their ownership of rapidly appreciating real estate assets offers them the chance to retire in comfort (should property remain the favoured investment in this country) at a time when social welfare cannot offer the same. This particular cross-section of owners and professionals hold on to an image of an Aotearoa based on wealth and success for those with the desire to take it, even as across the country as a whole, outcomes rapidly become ever more unequal and wealth ever more concentrated.

For the majority of people in New Zealand, including even some belonging to this particularly fortunate demographic, the opportunities of increased financial extraction have not made up for the hollowing-out of public services and the erosion of work protections. Poor working people in Aotearoa have faced notably worse outcomes across the board for health, housing, and other formerly publicly steered institutions since the realignment of state interests in favour of finance (Barnett & Bagshaw, 2020). This country's "essential workers" have for decades paid the price for the incredible opportunities handed to those professionals and owners in the wake of New Zealand's economic reorganization. This chapter is thus an analysis of the relationships between and within two particular groups at the urban periphery in Kumeu-Huapai and elsewhere: the professionals who have made their careers and their fortunes from the dramatic transformation of the landscape and strongly identify their own success with ideas of "equality of opportunity", and the working population of the city that build the houses, harvest the crops, serve customers, and otherwise ensure the existence of the urban fringe, yet are not necessarily granted membership in the community or even considered to be related to this suburban landscape. Class, along with other forms of social categorization, are often difficult to discuss or even discern in New Zealand, but the rapidly urbanizing periphery is an inflection point for the unequal relationships that dominate life in Aotearoa, and thus, formerly rural towns at the periphery of urban centres can offer unique insights into how divisions of Class and Race are formed and expressed.

## The material basis of class at the colonial urban periphery

For more than 150 years, the landscape of the Kumeu River Valley has been repeatedly transformed for the purposes of creating profit for financiers and landowners, many of them based in Auckland and in other Antipodean urban centres. In seeking such profitability after the forests of the valley were first cut down, owners turned to the marketing of land parcels to professionals and new arrivals in the country, selling the swamps of the valley on the imagery of bucolic splendour. Speculative financial interests became less relevant under the state-led initiatives of the mid-century, but never entirely left the valley. With the suburbanization of the periphery, this attitude from capitalists towards land at the edge of the city is once again the dominant political economic concern, here at the place where the expansion of urban life meets the collapse of rural production. As has been detailed throughout this book, the ideal resident of a new suburban home at the periphery of the growing city is a salaried professional with the means to afford a mortgage for a 700 $m^2$ house on 800 $m^2$ of land (yet not enough to simply buy property in the city), the cultural inclinations towards colonial land ownership that makes them useful targets for advertising material, and the motivation to be seeking out a permanent home in the first place. It is therefore unsurprising that the residents of Kumeu-Huapai, according to the last full census in 2018, are married at a higher rate than the general population (59 per cent vs 48 per cent), have a higher European portion of the population (82 per cent vs 70 per cent), and are more likely to be employed full time (62 per cent vs 50 per cent, and in fact, more than 75 per cent of men in Kumeu-Huapai work full-time), with a higher proportion of these in managerial positions (25 per cent vs 14 per cent). Kumeu-Huapai's median income is $47K NZD while the nation median sits at just below $32K, and nearly a third of residents earn over $70k, compared to 17 per cent across New Zealand; 85 per cent of them drive to work, either in a private vehicle or a company car (Stats NZ, 2020c). The people who make up the rapidly growing population of Kumeu-Huapai today are not Aotearoa's richest or the most significant beneficiaries of the FIRE economy. These truly wealthy people tend to be found in Auckland's increasingly exclusive inner suburbs or, as with the wealthiest of former horticulturalists who sold their land to developers, in rural retreats and estates. What these new residents of the river valley do represent is that particular demographic of professional salaried workers and managers, as imagined by Bourdieu (2005, p. 26), a source of demand, even something of a captive market due to the massive increases in house prices in the past two decade, that developers are seeking to capture by transforming the urban periphery.

The new residents are themselves active agents in capital accumulation on their own terms, albeit on a much smaller scale than the groups that made the river valley what it is today. While every aspect of the "foundations of petit-bourgeois suffering" is certainly present in the lives of these new peri-urban residents and commuters, finding themselves with million-and-half-dollar mortgages and spending hours a day in their cars driving to work, school, and their social lives across the city, the suburbanization of the periphery has enabled these professionals to at least become property owners at a time when real estate is a far more important determinant of middle-class wealth and comfort than a professional-managerial income. Just as in the rest of the Anglosphere, real estate asset prices in New Zealand have risen far beyond the rate of professional salaries. At the turn of the new millennium, a median-value house in Auckland cost some four times the median household income, and by late 2021, this ratio had tripled, meaning the same property cost (as noted back in Chapter 4) 12 times the median income (Interest NZ, 2022). Such a dramatic appreciation makes property ownership by far and away the most crucial form of wealth for any working person hoping for "upward mobility" today. The rise in the value of these assets – mostly one's family home – also offers the opportunity for financial leveraging, relying on ever-rising asset values and cheap credit to acquire more property using their house as collateral. This phenomenon, as seen in my discussion of valuations of land back in Chapter 3, has created a sliver of 8 per cent of homeowners carrying some 40 per cent of the nation's mortgage debt (Rehm & Yang, 2020). This particular source of returns, speculating on the value of a family home, has increasingly become the sole guarantor of wealth for the middle class, or more precisely, the "middle three quintiles" of people living in the world's wealthiest nations (Wolff, 2013). Out in the future suburbs of Auckland like Kumeu-Huapai, this search for real estate assets that offer good financial prospects can be seen in the resident selling their home to move a few streets over after two years, the advertisements for homes for sale emphasising the possibility of further development and subdivision of the sections actually being sold, and indeed in the unprecedented rise in the price of a house in these peripheral areas over the last decade. While banks and other larger financial interests demand an endless extension of residential suburbia for reasons of profitability, individual prospective homeowners see the potential for their own personal financial strategies to bear fruit, utilizing their salaries or existing properties elsewhere in Auckland to rapidly trade up.

These concerns dominate the interests of homeowners in peri-urban California through the same incentives that define the strategies of Kumeu-Huapai residents. Behind Fresno's legendarily corrupt practices of enabling sprawl and encouraging maximally profitable development in its poorest areas stands the sentiments not only of those involved openly and directly in property development – the property developer with a custom REZONE license plate on his vehicle, the Council members bypassing legislation to push through suburban gentrification schemes – but also of professional-managerial class homeowners in the City's northeastern expanse. On top of a broader centuries-long history of citizen-led segregation (Zuk, 2013, p. 31), private housing groups and their political representatives have fought against any restrictions on development forms in favour of affordability over profit, warning it would "drive away all the middle class and upper middle class people" (p. 84) and thus that crucial source of demand driving the rise of house prices at the periphery. When residents of Fresno's southwest do make requests to the City Council for such measures, the response is confrontational and particularly dismissive, the City claiming that these petitioners don't know what they're asking for. One council member said of them "They're advocating on a memory. What do they know about the neighbourhood? It's a different place now" (p. 117), while another dismissed a set of suggested attempts to control financialized development as, of all the descriptors to use, "Berkeley Bologna" (*ibid*). These representatives – and the homeowners groups that back them – are fully aware of the relationships between suburban developers, banks, prospective property owners, and local government that make rising house and land values a reality and spend considerable energy protecting such an arrangement.

Fresno is an internationally significant case of the power of property interests over urban development, an extreme example of segregation and profit-making of some city-dwellers through the displacement of externalities onto poor residents. Yet, this same pattern of interests is just as present among property owners in outer Melbourne, even if they are expressed in radically different terms under Victoria's own political context. As the City's frontier moved from Sunbury to Gisborne and the wider Macedon Ranges area, the Shire's planning authorities set out plans to allow for commercial development and tourism while protecting local agriculture. In doing so, they found a significant source of opposition from local landowners demanding that zoning changes – from rural conservation and residential – to rural production and commercial – be forestalled. Macedon Ranges Shire's planning and environment director Angela Hughes noted that landowners and developers hoped to keep a status quo arrangement to allow continued speculation on this soon-to-be suburban rural land, rather than allowing it to be earmarked for far less profitable horticultural or commercial purposes. "The strategy recommends the shire's rural landscape be protected, so it will not be turned into ad-hoc housing settlements in the future" (Preiss, 2021). Closer in, towards Sunbury and the rest of outer Melbourne where suburban development has already profitably transformed the formerly landscape and professionals buy detached houses an hour's drive from their work and lives "simply because that is the cheapest dwelling available" (Donaldson, 2019), homeowners are concerned with maintaining financial profitability of their assets through the more familiar activity of preventing denser developments overtaking their suburbs in much the same way as in Fresno or Auckland's periphery. Their goal in all three cases is maintaining the suburban character of neighbourhoods for the purpose of protecting the social stratification and aesthetic desirability (Cook et al., 2013) that attracts moneyed professionals seeking to buy an affordable home (and in turn capitalize on it as an asset) in a dramatically financialized housing market.

For property owners in Auckland's Future Urban Zones, in the Shires at the frontier of outer Melbourne, at the margins of the cities of the California Central Valley, and in a thousand other places where capital from urban centres seeks profitable reinvestment in urbanization, the appreciation of property values acts as a powerful incentive to act in favour of abstract financial accumulation over more material local concerns. As shown above, peri-urban landowners can act as social and political bloc and demonstrate an understanding of how they can extract unmatched profitability from their suburban homes and lifestyle blocks. As one local official in Kumeu-Huapai put it, "they're not shy about rolling up to MPs offices, working with developers, this is like winning the lotto … there's a lot of money at stake". For this reason, peri-urban homeowners constitute a noticeably distinct class in the rapidly growing satellite settlements like Kumeu-Huapai. More than 40 years ago, Barbara Ehrenreich and John Ehrenreich (1977) sought to separate out a subsection of the working population in mid-century United States to find the common interests that drove the actually existing American "middle class". Starting from the work of Marxist theorists who had long been troubled by an "unbridgeable objective class distinction" (p. 10) between well-educated salaried professionals and the rest of their working compatriots while still not conforming to most definitions of the bourgeoisie or owner class, Ehrenreich and Ehrenreich defined an economic base and social superstructure for what they called the "Professional-Managerial Class" (p. 11). These people work on the basis of salaries as "mental workers" or "skilled workers" across all the sectors of working life, in, as the name suggests, the role of intra-industry career professionals or as the managers of other workers. They are thus notable for being working people with a significant interest in "the reproduction of capitalist culture and class relationships" which has given them such exalted positions (p. 15).

Much more recently, in 2014, David Graeber looked back on this definition offered by Ehrenreich and Ehrenreich as part of a discussion of the role of social class in political activism and scholarship and made one significant update. In the decades since the concept of the Professional Managerial Class has been coined:

> the upper echelons essentially shifted their allegiances and realigned with the financial classes … financial elites and corporate bureaucrats essentially merged: the two classes began to intermarry; their careers tended to move back and forth between the different sectors; they came to speak the same language, share the same tastes, and see the world in identical terms
>
> (Graeber, 2014, p. 74)

He held that members of the professional-managerial class had once been aligned with what made their individual professional position lucrative, be that the profitability of their corporations' activities or technical advances in their field. Since the global resurgence of financial control over life, however, they had collectively become ever more reliant on their relationship with financial extraction to sustain their careers and their lives. The way in which financialized residential development is experienced by prospective buyers (rather than developers) is a particularly stark expression of this dynamic. The new residents of Kumeu-Huapai, like homeowners in myriad other suburban frontiers of urban financial centres across the planet, earn more money and sit higher on the professional hierarchy than the national average, but no matter where they work, their most crucial material interest is gaining property on mortgage terms they can survive and then do all they can to ensure the speculative value of their new family home continues to rise.

If the professional-managerial class is distinguished at Auckland's urban periphery by their leveraging of salaries to acquire property in the hope of finding sanctuary in financial gain, the rest of the working populace are defined by their lack of entry into local property ownership and a corresponding lack of membership in community organization. During a meeting to prepare for the upcoming Kumeu Show, a member of the Kumeu Agricultural and Horticultural Society remarked to me that the previous organizer of the local Neighbourhood Watch had refused to allow any tenants living in the river valley to join, referring to them all as "transients". At another time during my fieldwork, I attended a couple of well-publicized crisis meetings on the issues faced by the rapidly growing town. To set the tone for the urgency of these discussions, one official began by hurriedly telling the audience "We are all perfectly aware of how bad things are … we already know these things" before the same familiar problems (of traffic, of sprawl, and of the lack of supporting infrastructure) could be immediately bought up. The questions at these events, which soon become long-running conversations and eventually arguments, were dominated by three groups of people – local property development professionals concerned with their image in the face of the externalities of their success, older long-time residents who generally knew the local officials presenting the facts of the issues personally and had decades-old quarrels with them, and a handful of couples with young children representing the new contingent of owners concerned about the conditions their peri-urban move had placed them in. As real estate professionals assured those in attendance that unplanned developments would be avoided, older residents demanded a particular local official unblock them on Facebook or publicly apologize for calling the town of Glorit a collection of overgrown paddocks, and occasionally a newer resident would ask that same unanswerable question about when infrastructure would come to Kumeu-Huapai, no representative seemed to exist for people who lived and worked in the river valley but were not local property owners.

One perspective of life in the area for those who lacked the cash to buy into the local housing bubble that has since stuck with me comes from a journalist's interview with a man who worked in one of Kumeu-Huapai's liquor stores. "I rent a room in Hillsborough for $150 per week, but a room for rent in Kumeu is advertised between $250 and $300 per week" (Lovell, 2019). The journalist goes on to note that Hillsborough is 30 km away, "on the other side of Auckland". The liquor store worker drives this distance, across the heart of urban Auckland, to arrive at its western periphery every day, and when his work is done, he leaves. He did want to live in Kumeu-Huapai, he told the journalist, there just weren't any places he could afford to rent. Through both intentional exclusion and via the more abstract demands of financial profitability from real estate that drives so many to the periphery, working people who ensure the day-to-day functioning of Kumeu-Huapai are separated from the social lives of their professional compatriots who have managed to enter into debt relationships and become property owners.

The financialization of real estate in the Kumeu River Valley is a powerful basis of class interests among property owners at this frontier of Auckland's urban sprawl. Rapidly appreciating property prices across urban centres combined with the uniquely massive potential profitability of these new suburban developments has created a lucrative (albeit extremely unstable) relationship between the development sector and professional-managerial workers with sufficient income to receive a mortgage and thus hopefully ride a wave of inflated asset prices to an escape from an increasingly precarious working life. This group of exurban lifestyle residents and new suburban homeowners are very much aware of the parameters of this relationship and actively engage in attempts to protect it, just as their compatriots do in Victoria, Australia, and California, USA. The cost of such financially oriented social strategies is an unavoidable alienation of the rest of the working populace from these communities. Housing becomes difficult to find for any of these less-well-paid yet nonetheless very vital service providers, both due to attempts to maintain class stratification and housing forms that attract highly paid professionals, and from the massive speculative appreciation in the cost of housing itself. While land bankers and developers are certainly the largest and most visible practitioners of capital accumulation from the rural land at Auckland's fringe, their customers also rely on the same process of financial extraction for their own gain, and the costs are borne by the rest of the working people in the river valley.

## "When I moved here it was a lovely country town" intra-class conflict, allegiance and race in the suburbanization of rural Kumeu-Huapai

As much as the land-owning residents of Kumeu-Huapai do constitute some very clear shared economic and social commonalities that constitute them as a distinct class, this does not imply they act uniformly and without internal conflict. As Ehrenreich and Ehrenreich noted, "class is analytical abstraction, a way of putting some order into an otherwise bewildering array of individual and group characteristics and relationships" (1977, p. 13) rather than some crystalized and unimpeachable model of reality as it is truly experienced. In fact, one of the loudest and most blatant conflicts I noticed during my fieldwork was a particularly sharp split between two groups within this same propertied professional-managerial class. The population of the town doubled between 2013 and 2018, and judging by the Stats NZ predictions noted back in Chapter 2, has most likely already doubled again, yet many long-time residents maintain a sizeable presence over local institutions such as community action groups and the Agricultural and Horticultural Society. Many of these more established residents moved to the Kumeu River Valley in the decades when the area was ideal for city-based professionals to buy several acres of rural lifestyle living or had grown up when agricultural production still dominated the economy, before enthusiastically, grudgingly, or bitterly transitioning their life, work, and land to the new demands of the financialized periphery. As developers seek more and more dramatic transformations of Auckland's northwest to accommodate suburban development at maximal profitability, and the more recent faction of professional-managerial residents attempt to impose their own visions of the ideal expressions of their social class onto community life, these older property owners find themselves, in turn, struggling to enforce their understanding of the landscape onto parties that imagine entirely different futures for the land and people of Kumeu-Huapai. Even while they operate under shared financial incentives, significant gulfs in allegiances to particular ideas of rural and suburban life drive Kumeu-Huapai's professional-managerial residents to conflict over what the river valley should become and who should live there.

At the end of an interview with a Kumeu-Huapai resident whose home had been terribly damaged by the floods that swept through town at the end of August 2021, the very event to which most of Chapter 8 is dedicated, the conversation turned quite rapidly when the interlocutor inquired in more detail about my line of research. The moment the topic of the impact of urbanization on work and life in the area came up, they stepped back from their assessment of the damage to their house, the piles of furniture and whiteware now awaiting to be hurled into a skip, the chunks of the interior wall cut out at waist level pending complete replacement, the vehicles now officially written off by the mere touch of floodwater above a certain margin, and they began to address a far slower transformation than that brought by the once-in-a-generation flood. "This used to be a lovely country town … we'd ride horses down this road", until traffic began to dramatically increase a decade ago. This activity was no longer possible, and so their entire relationship with the town had changed.

This homeowner belonged to the contingent who had moved to the Kumeu River Valley some 30 years ago in the area's first phase of popularity as a lifestyle destination for Auckland residents. As suburban frontiers expanded, and developers began to sell the rural aesthetic and strong investment prospects of the area to more and more urban professionals, this very particular balance between proximity to the city (by car) and the image of a secluded rural landscape ended. I discussed in previous chapters how the very successes of peri-urban development in Auckland's northwest have in turn required a shift in advertising to accommodate the unavoidable realities of a much less rural Kumeu-Huapai. A very real change has taken place in aesthetic terms, from the resident who was sold on the idea of engaging in agricultural cultivation as a hobby in between their workdays in the city to the new homeowner being informed they should be buying a large house that covers most of the property's land plot because they won't have time to maintain their lawn. Beyond the visual transformation, however, is a marked shift in the practices and activities considered suitable for a homeowner in the area. Once-popular social clubs and community organizations are struggling to take on new members (see Figure 6.2) and new centres of social life are gaining power as these older networks fade. What it means to be a property-owning resident of the river valley, the social relationships, and practices that define a membership to some ideal of peri-urban life is changing at an accelerating pace.

*Figure 6.2* Notice from the Kumeu-Huapai Lions calling for new members to save the club from collapse after a half-century of operation. Photo taken by the author.

Back in Chapter 2, I presented Ferguson's historical approach to capitalist development in the Zambian Copperbelt – a focus on clusters of variation produced by economic relationships, a barometric chart of accumulation and abandonment rather than a teleological line from "savagery" to "western civilization" – to reassess narratives of the Kumeu River Valley's relationship with financial accumulation since its colonization by the British Empire. Here, another aspect of his theoretical approach is useful for understanding the intra-class disjuncture among Kumeu-Huapai landowners: Style as a micropolitical expression of allegiance (Ferguson, 1999, p. 92). Ferguson noted that the boom of the mining industry in the Copperbelt over the mid-century had allowed many professionals to reject local ties to family and village in favour of an adherence to a cosmopolitan social life. These upwardly mobile workers, living in a nation that was gaining international recognition for its rapid "industrialization" or "development", dressed, spoke, ate, and drank like the "modern" western-oriented global citizens they were sure to become. When the global restoration of finance over governance drastically curtailed this local accumulation of wealth and shattered the Copperbelt's urban economy, it started to feel for these professionals like history was going backwards, that modernity had abandoned them (p. 11). While some held on to a cosmopolitan allegiance in their social life, continuing to embody the practices and behaviours of someone destined to be "out there", many urban professionals made an intentional effort to learn how to become a local of the countryside, to rebuild ties with rural kin and act in ways that would please their elders in order to start a new life on the land (p. 112). A young man drinks the locally brewed dark beer, hangs around older men to learn their mannerisms, and abandons a linen suit and tie for a self-consciously "traditional" outfit that indicates to any passers-by who his allegiances sit with.

The point of Ferguson's observation was that these performances were both intentional moves to invest one's life into a micropolitically advantageous position and simultaneously the kind of ongoing sedimentation of behaviours (conscious and unconscious) that Bourdieu (1987) would describe as being formative of social class in bourgeoisie life. To engage in agricultural production as a hobby when you aren't at work in the city, to own and ride horses down the roads of the town on the weekend, and to take part in meetings of the Horticultural Society or the Lions Club are a conscious expression of what you believe the land of the urban periphery is for, and a continuous process of socialization within a group of peers that possess a similar set of inclinations. In trying to piece together some clear explanatory perspective on the "bewildering array of individual and group characteristics and relationships" that make up social relations among property owners at the urban periphery, the rise and fall of these practices and groups is one particularly clear heuristic from which I seek to draw the dividing lines in this intra-class conflict.

If the roads are no longer suitable for weekend horse riding and the land is no longer valued for agricultural production even as a pastime, how do these new professional-managerial residents of Kumeu-Huapai make use of their space? One useful case study of changing inclinations at the periphery is the decline of the Kumeu Agricultural and Horticultural Society's annual Show, first discussed in Chapter 1, compared to the rapid growth of another yearly event held on the very same grounds, The Kumeu Classic Car and Hot Rod Festival. The former has been with Kumeu-Huapai since the first few families that had bought into the Huapai Estate's rosy promises of bucolic idyll successfully converted the swampy deforested valley into productive land, the first show occurring in 1921 (Dunsford, 2002, p. 65) and only has ever been cancelled twice in the past century. The first was in reaction to the beginning of the war in the Pacific following the attack on Pearl Harbour in 1942, and the second was in the final weeks of my fieldwork in 2022, when a sudden spike in COVID-19 cases led to restrictions on mass gatherings. It survived through the apathy and poverty of the great depression (p. 137), it thrived under Kumeu-Huapai's status as both productive agricultural site and go-to rural destination for Aucklanders in the 60s and 70s (Kumeu Agricultural and Horticultural Society, 2022), and it successfully made the transition from a rural trade show to a two-day local fair by the new millennium, after agriculture in the river valley collapsed in the 90s. By the time of my fieldwork in 2021–2022, the show and its society continued to be financially solvent.

Yet, careful management of accounts could not protect the society from the steady social transformation occurring at the periphery. When I attended meetings of the society as they planned the upcoming annual show, I was asked regularly if I could become a member, if I could convince my friends to take an interest in the operation of the society, if my research could act to publicize the importance of such societies. The members who attended and engaged in setting up the annual show were a close cohort of long-time residents aged between 40 and 80 years old, drawn from that older subset of professional-managerial landowners who had moved here to participate in the practices of rural life, or had grown up in the Kumeu River Valley and remembered its time as a productive hub. As I described all the way back in Chapter 1, the end of this era also meant the collapse of the economic basis for the show's relationship with its biggest partners and customers. Only a single produce stand remains, the brand-new tractors are mostly for children to climb on, and the rows of utility vehicles to be peered at by passing couples. An event that was once "the biggest one-day show in the southern hemisphere" and a "place to make deals" now survives as a (still very prominent) local fair with a collection of rural-themed activities that offer "an opportunity to reconnect" the suburban attendees with agrarian life. The inclinations and practices that would bring in new members into the society and the show no longer exist, and thus the social life of the Agricultural and Horticultural Society faces jeopardy long before any financial matters threaten its existence.

In marked contrast, the organizers of The Kumeu Classic Car & Hot Rod Festival seem very concerned chiefly with the rapid pace at which the bounds of their event began to exceed their ability to host it. "We created a monster". From its birth in the 1990s, the Hot Rod Festival became only ever more disruptive every year it was held, holding up traffic across the northwest and requiring the assistance of local authorities. The number of attendees has stabilized at around 30,000–35,000, something like double the tickets sold to the Kumeu Show today, although this is still little more than half of the supposed peak of the agricultural event's prominence in the 1970s. The day before the event begins, long-time attendees and their families camp out together on the Agricultural and Horticultural Society's showgrounds and drink well into the night. On Saturday, classic cars pour in from across the nation, while locals sit on folding chairs on the side of the road to watch them glide past. The showgrounds are filled with ranked masses (see Figure 6.3) of American, Japanese, and European cars of every vintage from early model Fords to 1990s Toyotas. In a gravel section normally used as a parking lot for neighbouring businesses, car parts and tools are laid out for perusal on crowded tables or simply laid out on a blanket in front of the owner's tent or car (see Figure 6.4). Like the Kumeu Show, the Hot Rod Festival hired a number of fairground attractions (Figure 6.5) and opened space for merchandize stalls to make their trade.

The standout difference between the two events was not the numbers of people attending, but the interactions within a core of festival-goers, some of whom had been meeting at this place every year for more than two decades. "We see people here we normally only ever see here" remarked one woman, and a man sitting next to her told me with a grin "missed weddings, missed funerals, never missed this".

*Figure 6.3* Rows of cars bought along by attendees to be shown off or occasionally sold, filling up the showgrounds. Photo taken by the author.

*Figure 6.4* Examples of the ways in which automotive hardware and other second-hand goods were sold during the show. Photo taken by the author.

Many of those trading car parts and other merchandize also saw the money-making aspect as an entirely secondary concern to socializing: "we just put out a few things for swap so we can set up our tents and get shit-faced here". The ongoing success of the car show, especially in the face of the disruptions wrought by COVID-19 in

*Figure 6.5* A stunt rider takes flight above the showgrounds. Photo taken by the author.

the past two years, is driven by the continuous reinforcement of these social ties, providing the event organizers with a supply of volunteer labour (every person working the event was a family friend of the organizers), word-of-mouth advertising, commercial sponsors, and technical expertise, the exact things the Horticultural and Agricultural Society has been steadily losing.

The Classic Car and Hot Rod Festival is undeniably a major local event and an important part of life in the area, but it does not sit well with many long-time residents. One of the organizers told me "we put Kumeu on the map", but when I asked which local businesses or figures tended to get involved, assuming they would have recognized the money and connections to be made, he laughed and said "oh they want nothing to do with us". Rather than carrying sponsorships from local concerns, the organizer claimed, they were forced to go international, flying the logos of Australian and especially American companies at the festival. Whether this was actually due to some embargo from local elite on the event is entirely unclear, beyond the fact that whenever I brought up the Agricultural and Horticultural Society with the runners of the Hot Rod show, I would be told that the Kumeu Show was a few years from collapsing, or when speaking about the latter to the former, that those who ran the Hot Rod show were an impediment or a nuisance to the operation of the showgrounds. Regardless of the personal basis for the choice, the Hot Rod Festival was notably for an overt "international americana" aesthetic, rather than the intentionally parochial representations of mid-century Kaipara that dominated the Kumeu Show. Prominent displays of American flags and references to American automotive arcana such as Route 66 and the Daytona Speedway (see Figure 6.6),

*Figure 6.6* Examples of Americana at the Festival. Photo taken by the author.

and a booth set up by a group of US Marine Corps reenactors, all indicated an interest in the consumption of a global American culture. One of the organizers of the Hot Rod Festival, a long-time resident with experience working in advertising, believes this to be an especially important set of symbols and inclinations that they should exploit. From over-the-top publicity stunts, "we had 14 elvises jumping out of a plane, we tried to set a [*world*] record for elvises" to a second show marketed at women and families called "A Very Vintage Day Out" that incorporates mid-century American fashion and music, this organizer believes utilizing such imagery will only become more important for the popularity of the Hot Rod Festival. "Everyone's into it, they love this stuff". Rather than attempting to reconnect with the Kumeu River Valley's own rural past, this event finds financial success and the creation of lasting social relations by appealing to a set of inclinations rooted in the imagery of worldwide Americana.

This divide in social life between older and newer property owners has implications for how race is used as a marker for social class at the urban periphery. Particularly among older residents, Kumeu-Huapai is an explicitly Anglo project into which Dalmatian settlers eventually integrated by also becoming English-speaking landowners. The supposed historic absence of Tangata Whenua – "there's no Māoris here" – has left mid-century cultivators of the river valley as the "original" inhabitants in their own eyes. Older interlocutors expressed opposition towards attempts by iwi to assert their sovereignty over colonial land in any way, as well as suspicion towards the "Asians" or occasionally the older category "ethnics" who made up a small but noticeable proportion of the new professional-managerial residents of Kumeu-Huapai. These older residents, in a dynamic perhaps familiar to readers, occasionally pointed to Asian property ownership or to events such as the removal of golliwog dolls or the ceasing of the publication of certain visibly racist Dr. Suess books or Tintin comics as troubling signs of a changing middle-class culture.

Property development professionals, looking at the all-important demand for suburban housing in their business model, take a different, if not entirely opposing view. Real estate agents and others in the property sector expressed a satisfaction with the slightly increased diversity among "all these young families" that made up the new suburban residents. One manager, clearly still uncomfortable with this visible change in racial significations of class, said that they were happy the town was "still quite a friendly place" despite "people from the city moving here". Similarly, a comment between organizers of the Hot Rod Festival about the culture of their show led to one remarking on the area being "very white … Kumeu is a very white place" and the other telling me "but the Indian boys do love cars too, we're getting more of them". These sentiments are very relative; Kumeu-Huapai, as noted earlier in this chapter, is still overwhelmingly occupied by wealthier-than-average European residents, with the most notable demographic change being a rise in the proportion of Asian residents from just above 3 per cent to a little under 15 per cent, as well the combined Latin American, African, and Middle Eastern population rising from 0.0 per cent of residents in 2006 to 1.7 per cent by 2018 (Stats NZ, 2020c). This small change, however, has highlighted a noticeable difference between the inclinations of older established landowners and those less invested in their understanding of Kumeu-Huapai. For the former contingent, the river valley's social life was built on the same strict racial order that put immigrants from the Balkans in the same category as those from Syria and China (Bozic-Vrbancic, 2006) in order to enshrine the rights of Anglo colonists. For the latter, membership through property ownership and professional-managerial employment allows a (limited) relaxation of this stratification.

Issues of race and membership articulate very differently between some older and newer residents when it comes to schools, however. Among the high-priority items demanded by Kumeu-Huapai's residents – public transport, a highway by-pass, etcetera – a new school always featured prominently. When local National MP Chris made public demands for action on infrastructure and held a protest along the highway (see Figure 6.7), the funding and construction of a new local high school was one of his four points for improvement. The two nearest schools, Kaipara College to the west and Massey to the east, were either too full to accommodate the growth of Kumeu-Huapai, or not suitable for the children of the new suburban residents. As Chris Penk himself put it, a few years ago these people didn't want to send their kids to Kaipara College, "but their reputation is much higher now". Yet, Kaipara's new school zone will exclude Kumeu due to capacity issues, so "now only Massey is available … some people in this area feel that [Massey] is a different part of town". Massey – the school and the area – carries a reputation for Kumeu-Huapai residents as being poorer, less white, and more dangerous than they would like. Some older residents, however, completely disagree with this point of view. "They're not using the local schools … a bit too brown for them" one long-time local official noted, but then said that older families continued to support Kaipara and Massey "they have a different worldview". A member of a local community action group told me "It's disguised, well its actually quite blatant racism … they want their kids to go to the North Shore with White and Asian kids". While older colonial structures of racial stratification in social life are now

*Figure 6.7* National MP Chris Penk's protest on State Highway 16 against the lack of infrastructure accompanying development. Photo taken by the author.

less important in Kumeu-Huapai, the new professional-managerial residents of the river valley are very concerned – far more so than those who moved here decades before – with ensuring their children are socialized in the way that best promotes membership into this very same professional-managerial contingent.

While operating on very similar material interests and interacting within many of the same community groups, the landowning residents of Kumeu-Huapai are still divided by conflicting social inclinations and allegiances created by the rapid suburbanization of the periphery. Long-time residents who had been born in the Kumeu River Valley or moved to the area seeking a rural lifestyle decades ago continue to adhere to behaviours and social relations that value at least the aesthetic of a parochial and agriculturally productive town in between Auckland and the Kaipara coast. For the newer suburban arrivals, however, these inclinations hold far less relevance. Successful events and organizations do not draw from memories and dreams of the river valley's provincial past, but from imagery that connects participants to global – particularly American – symbols of nostalgia. Parents sending their children to school are less concerned that their own children grow up alongside other children born in the North-West than they are with ensuring they are educated and socialized in a manner that reinforces lines of social class. As Kumeu-Huapai's population continues to grow exponentially from a town of little more than a thousand to that of a suburban frontier of Auckland City with tens of thousands of residents, this latter set of practices and worldviews promises to overtake the former, completing the transformation from a locally centred social life to one operating in relation to the urban centre.

## Who is the financialized urban periphery for?

"This area has always been kind to people wearing Blue". Chris Penk, National MP for Kaipara-ki-Mahurangi, is telling me that this region has supported his party since it was formed. Three-and-a-half term New Zealand Prime Minister John Key entered politics as the local MP for this area, and he remained the representative for the electorate until his retirement and replacement by Chris. At agricultural shows, community fairs, and school events, I grew up hearing John Key's oddly unplaceable voice over tinny PA systems making statements I can barely remember beyond some allusion to the unique opportunities offered by this country. Born the same year as my parents, John Key was famously raised by a single mother living in state accommodation, before graduating from university with a commerce degree and quickly finding himself in the rapidly expanding field of foreign exchanges. Key ascended from local dealings by his early 30s to begin working under the auspices of Bank of America's investment arm Merrill Lynch in Singapore and then London. Like many of his place and time, he was offered housing and education under state-led initiatives despite being born without wealth and began his career at a time when the liberalization of New Zealand's (and the world's) economy offered unparalleled extractive opportunities for those with the capacity to act on them. Once in politics, he called the very idea of housing being constructed and sold under terms not dictated by market interests "economic vandalism" (Thompson,

2006) and spearheaded efforts to accelerate the privatization of remaining public assets (Small, 2012). John Key's trajectory closely matches the lives of many older professionals I interviewed, though Key is of course significantly richer today than most of these interlocutors could hope to be. For this relatively small number of the nation's population, the state gave them the ability to capitalize (literally) on its own annexation by financial interests from the 1980s onwards, and finance-led suburban development in rural Auckland has given them a career, an astonishingly lucrative investment vehicle, and a path to retirement in comfort. From their perspective as a voting bloc, the National Party's policies of a moderate social conservatism matched with an open embrace of continued financialization in every aspect of work and life have been unquestionably successful, and so Chris Penk remained safely in his seat even under the wave of support for the opposing Labour party in the past five years. For everyone else, from the people who work to provide the service they rely on, to older working people who never managed to acquire this wealth, to even those other relatively wealthy professionals who benefit from the process accumulation yet are stunned by the failures of development in the area, this incredible financial success brings costs that only continue to mount.

Rates of homeownership and wealth have risen in Kumeu-Huapai with its suburbanization, yet the opposite has simultaneously occurred in the town's own rural frontiers, as land becomes most valuable as a speculative asset in the path of future expansion. As noted earlier, some two-thirds of the residents of Kumeu-Huapai are homeowners, while another 13 per cent hold their property in a family trust, meaning overall only a fifth of the town's population are tenants (Stats NZ, 2020c). In the two rural areas of the river valley under the census, labelled simply Kumeu Rural West and East, property ownership fell from 50 per cent to about 40 per cent between 2006 and 2013 in Kumeu Rural West (Stats NZ, 2020b), and all the way down to 36 per cent in Rural East (Stats NZ, 2020a). At the same time, a very high number of residents −30 per cent – now hold their properties in the rural expanses of the Kumeu River Valley in family trusts. The average annual income for these areas sits at $38K NZD (West) and $35K NZD (East), slightly above the national average and well below Kumeu-Huapai's number of $47K. Looking out the window of their home on a low-lying and flood-prone lifestyle block east of the town, one resident said "not many owner-occupiers around here, it's mostly Chinese or Korean land bankers. Don't know why they decided to buy here, this is probably going to be industrial. That might make it cheaper though". The process of transforming rural areas into a hinterland suitable only for development, described in Chapter 3, makes places like rural Kumeu into sites of financial speculation first, with the viability of all other land uses sitting in relation to this goal. As the new suburbs, built in more suitable parts of the river valley, are occupied by the wealthier city-based residents that can afford a mortgage, this hinterland becomes ever more associated with the abstract possibility of future development and less with work and life today.

While I have spent much of this chapter referring to the wealth and advantages gained by the members of the generation that were able to capitalize on the liberalization of New Zealand's economy, there was one rather poignant commonality between some older members of the community, whether they had moved here five or 50 years ago, regardless of their opinions on rural life. Remarking on the parking lots and fields covered in blankets laden with auto parts, the rows of gleaming classic cars with signs offering a negotiated sale, one of the Hot Rod Festival organizers told me that more and more long-time attendees were selling what they could, for one simple reason. “These baby boomers, they’re getting older, they have to start selling stuff … that’s why the swap meeting is getting bigger”. While some could rely on the unparalleled appreciation of property values to allow them to move out of the area and retire, as so many of my interlocutors intended, many others were faced with a significant challenge when it came to saving up for their old age. As the liquor store employee mentioned at the end of the previous section had noted, living costs here are much higher than average for the Auckland region, making it difficult for many to stay in the area if they don’t have the money. Even as some former horticulturalists spoke to me from the position of the comfortably retired, or at least that of not being required to work for a living, other older interlocutors admitted they needed to use the money gained from the subdivision of their land to start anew a different sector. The horticulturalist in Chapter 1 who had sought to juggle five collapsed orchards to stay afloat used the sale of their plot to a land banker as the basis for a green waste disposal business, rather than an escape from working life. Even for those people who did grow up at the right time and place to take advantage of the reorganization of the economy, finding a way to comfortably live out their later years in the northwest without significant property assets is by no means a simple task.

If the surrounding rural areas are being transformed into abstract speculative assets controlled by absentee owners, if elderly residents and working people cannot afford to live in the places they spend their days, then does the financialization of the urban periphery at least benefit the professional-managerial homeowners of Kumeu-Huapai itself? As Chris Penk pointed out in two separate comments, the voting residents of the area are steadfast supporters of the National Party and its commitment to a finance-focused approach to urbanization, while simultaneously “nine out of ten people who are in this office are here to complain about the state of the roads”. Even in the face of massive property value rises, the “foundations of petit-bourgeois misery” that define life at the new urban periphery are difficult to ignore. Chris’s protests and public statements demanding that public funds and resources be rapidly allocated to solve infrastructure issues created by urban sprawl struck many interlocutors as deeply ironic. “His government made these mistakes, and now he’s blaming labour for the mess” complained one landowner, while a local official burst out laughing when I told them about Penk’s protest along the highway, saying to me “he knows there’s [*only*] 150 Million allocated, and he knows why its delayed … he can wave his shovel all he likes”. They would later tell me “He writes letters to the ministers and then publishes them on Facebook … he told me it’s so he looks like he’s doing something”. Faced with the intractable conflict between the demand for maximal profitability from developers and land bankers against the constant complaints of homeowners old and new dissatisfied with life at the suburban periphery (yet still seeking to benefit financially from their acquisition of property), Chris Penk does his best to promise a resolution, and like the developers in the area, he looks to the state for the resources to resolve this contradiction.

Above all, then, the class of people who have benefited the most from the financialization of the urban periphery in the Kumeu River Valley are those who no longer have to live there. The younger Tony Nobilo, the part-owner of the land that became the Huapai Triangle, lives further out into rural Auckland now. Those long-time residents who can afford it are following suit, leveraging their properties to seek a new place to retreat from suburban life. The land bankers and the communities of city-dwellers that speculated on urban sprawl have had great success, even if some do become furious and demand law changes when they discover their land parcel has been zoned industrial or commercial rather than residential. As a financial product, the Kumeu River Valley has been very lucrative. As noted back in Chapter 1 in fact, Kumeu-Huapai was the most wildly profitable of any place in Aotearoa between 2014 and 2019. The costs of this financial success are at the level of daily life, of the inability for working people to find a place to live in the very town they are employed in, the shocking traffic, the non-existent public transport, the costly mortgages, and the general economic insecurity for people who are not large property owners. While wealthier local residents do actively seek to profit from the suburbanization of the river valley in their own way, they do so under the conditions created by developer and finance interests operating at regional and even national levels. Auckland's new urban periphery is most valued as a place to extract unmatched financial returns and is only a place for city-based professionals to find a rural lifestyle within the bound of this goal. Just as it was 150 years ago, the most crucial interests and decisions over the land of the Kumeu River Valley are made by a small number of financially motivated agents well removed from matters of daily life in the place they are transforming.

Kumeu-Huapai, like the many new suburban developments at the edge of Auckland and other growing cities across the planet, is very intentionally marketed to members of the urban professional-managerial class seeking an escape from city life or the opportunity to buy a large house at a price they can afford. For the property-owning residents who moved to the river valley decades ago or lived their whole lives here, these new homeowners present a challenge to the practices, inclinations, and social relations that had once defined life in Kumeu-Huapai for them, even if they share very close economic interests with their new neighbours. Similarly, the rapid growth of the town's population and the resulting slight increase in ethnic diversity are jarring for some older residents who still hold a strong allegiance to an explicitly Anglo Colonial concept of the relationship between markers of race and social class. Regardless of these conflicts and transformations in Kumeu-Huapai's social life among property owners, the people who actually work there every day are excluded from membership in the town both through intentional efforts to increase property values by reinforcing class lines, and from the high cost of living in Kumeu-Huapai as a result of these efforts. In the end, while the professional-managerial homeowners of the river valley seek the best outcomes for themselves through the rising values of their family home, they only do so in the context of the wider transformation wrought by individuals in the development sector, who engage in the most profitable forms of development possible by any means necessary. This same endless drive for greater forms of financial profitability for a few is drastically changing how urban residents relate to their food as well as their land. The next chapter shows how the peri-urban food system is dismantled not only by the demand for land to develop but also by the interests of individuals in the food production system itself.

**Bibliography**


Barnett, P., & Bagshaw, P. (2020). Neoliberalism: What it is, how it affects health and what to do about it. *New Zealand Medical Journal, 133*, 76–84.

Bourdieu, P. (1987). *Distinction: A social critique of the judgement of taste* (R. Nice, Trans.). Cambridge, MA: Harvard University Press.

Bourdieu, P. (2005). *The social structures of the economy*. Cambridge: Polity.

Bozic-Vrbancic, S. (2006). Mysterious degrees of whiteness: Stereotypes about Croats in colonial New Zealand. *Revija za sociologiju, 37*(3–4), 181–192.

Cook, N., Taylor, E., & Hurley, J. (2013). At home with strategic planning: Reconciling resident attachments to home with policies of residential densification. *Australian Planner, 50*(2), 130–137.

Donaldson, D. (2019). *Expanding urban sprawl: Melbourne is getting the mix wrong – The Mandarin.* Retrieved August 29 from https://www.themandarin.com.au/104204-expanding-urban-sprawl-melbourne-is-getting-the-mix-wrong/

Dunsford, D. (2002). *Doing it themselves: The story of Kumeu, Huapai and Taupaki.* Auckland: Kumeu District History Project.

Ehrenreich, B., & Ehrenreich, J. (1977). The professional-managerial class. *Radical America, 11*(2), 7–31.

Ferguson, J. (1999). *Expectations of modernity: Myths and meanings of urban life on the Zambian Copperbelt.* Oakland: University of California Press.

France, A., & Roberts, S. (2017). *Youth and social class: Enduring inequality in the United Kingdom, Australia and New Zealand.* Springer.

Graeber, D. (2014). Anthropology and the rise of the professional-managerial class. *HAU: Journal of Ethnographic Theory, 4*(3), 73–88.

Interest NZ. (2022). *Median Multiples*. Retrieved August 26 from https://www.interest.co.nz/property/house-price-income-multiples

Kelsey, J. (2015). *The FIRE Economy: New Zealand's Reckoning*. Wellington: Bridget Williams Books.

Kumeu Agricultural and Horticultural Society. (2022). *Kumeu Show – History*. Retrieved September 15 from https://kumeushowgrounds.com/kumeu-show/

Lovell, O. (2019). Where town meets country: Kumeu locals facing less rural future. *Stuff NZ*. Retrieved June 22 from https://www.stuff.co.nz/business/farming/113676968/where-town-meets-country-kumeu-locals-facing-less-rural-future

Nel, P. (2022). Mobility rules: Why New Zealanders oppose redistribution. *Kōtuitui: New Zealand Journal of Social Sciences Online, 17*(1), 24–43.

Preiss, B. (2021). *Showdown at the Macedon Ranges, Where Growth and Beauty Collide – The Age*. Retrieved August 29 from https://www.theage.com.au/national/victoria/showdown-at-the-macedon-ranges-where-growth-and-beauty-collide-20210923-p58u8q.html

Rashbrooke, M., Rashbrooke, G., & Chin, A. (2021). *Wealth inequality in New Zealand: An analysis of the 2014–15 and 2017–18 net worth modules in the Household Economic Survey*. Wellington: Victoria University of Wellington, Te Herenga Waka, Institute for Governance and Policy Studies.

Rehm, M., & Yang, Y. (2020). Betting on capital gains: Housing speculation in Auckland, New Zealand. *International Journal of Housing Markets and Analysis*, *14*(1), 72–96.

Small, V. (2012). *John Key Reveals Plan for Asset Sales – Stuff NZ*. Retrieved August 30 from https://www.stuff.co.nz/national/4582922/John-Key-reveals-plan-for-asset-sales

Stats NZ. (2020a). *2018 Census Place Summaries – Kumeu Rural East*. Retrieved September 15 from https://www.stats.govt.nz/tools/2018-census-place-summaries/kumeu-rural-east

Stats NZ. (2020b). *2018 Census Place Summaries – Kumeu Rural West*. Retrieved September 15 from https://www.stats.govt.nz/tools/2018-census-place-summaries/kumeu-rural-west

Stats NZ. (2020c). *Place Summaries: Kumeu-Huapai*. Stats NZ. Retrieved June 4 from https://www.stats.govt.nz/tools/2018-census-place-summaries/kumeu-huapai

Thompson, W. (2006). *Air Base Neighbours Fear Extra State Homes – New Zealand Herald*. Retrieved August 30 from https://www.nzherald.co.nz/nz/air-base-neighbours-fear-extra-state-homes/4HOWVT3M2EUOTCJAYUPFWP3GDM/

Wolff, E. N. (2013). The asset price meltdown, rising leverage, and the wealth of the middle class. *Journal of Economic Issues*, *47*(2), 333–342.

Zuk, M. Z. (2013). *Health equity in a new urbanist environment: Land use planning and community capacity building in Fresno, CA*. Berkeley: University of California.

# 7 Food sovereignty and peri-urban horticulture

Writing in response to the public rise in demands for the protection of productive land at the margin of growing cities seen over 2019, the chair of one of New Zealand's more prominent free market think tanks recommended that we simply build houses on more productive land and grow food in less productive areas:

> Rather than regulating whether land is used for housing and food, we can safely rely on markets to provide a solution: Where it is more valuable for housing, we should build on highly productive land and farm on less productive land. The latter may not be as easy to cultivate as the Pukekohe silt loam but it is a better choice than choking our cities and forcing up house prices to stave off a mythical food crisis.
>
> (Partridge, 2019)

*Figure 7.1* Horticulture plots sitting around the home of one of Kumeu-Huapai's extant growers. Photo taken by the author.

DOI: 10.4324/9781003400325-10

We can, he reminds us, import food, but we cannot import houses. This excerpt from an already very short article, written and posted as a matter of course to represent market interests in the urban sprawl debate at the time, gives us a dizzyingly inverted imagining of the world compared to the arguments that have been presented earlier in this book. In a vision that renders financial accumulation distinctly concrete and the idea of food production as a rarefied and abstract concept, land that currently provides a significant proportion of the country's fresh vegetables must be converted into suburban housing in order to generate the most profitable outcomes for a small number of people. As in the arguments made by development advocates back in Chapter 3, this demand for maximal profit-making is tied to a promise of easing the housing crisis. If any Auckland resident should want to see affordable housing (a mere hour-and-half drive each way from their workplace), the logic goes, they should embrace the suburbanization of places that currently serve to produce food for the city.

Attempts to calculate the relative value of peri-urban food production versus the release of land for exurban and suburban development have been a constant theme in the sources used throughout this book. Maps have been drawn up attempting to broadly interpret and categorize peri-urban land by its potential productive value, theories presented by professionals of all stripes on what places exactly should be considered worthy of protection, and arguments made by landowners and planners for whether legitimated transgressions of zoning law should be let through, all as multiple incarnations of central government proclaim their intention to streamline or clarify the entire conflict. Phrasing this argument as simply a need to protect "highly productive land", however, occludes the significant amount of work done by the process of urban sprawl to render horticulture less viable on land that visually appears to remain rural, as has been discussed earlier. Instead, calls from environmental scientists, small-scale food producers, and planners to deny the wishes of agents working in the property development sector are concerned with the related but distinct concepts of Food Security and Food Sovereignty. The former refers to the capacity for people to consistently get the food they need to stay alive and stay healthy, regardless of how many thousands of tonnes of food are being grown for export in their state (Birch & Keating, 2013, p. 5), and the latter broadly points to the ability of these same people to have control over their own food supply, rather than being reliant on the approval of other nations or groups to receive the food they need (p. 54). The ramifications of both terms will be teased out later in this chapter. It is in the criteria of these two concepts where it becomes apparent that not only is the "mythical food crisis" perhaps not such a ridiculous idea, but that a food crisis is already unfolding in Aotearoa.

## Starving in a land of plenty: food insecurity in globally significant foodbowls

The opening lines of an investigative story on food insecurity from Farah Hancock for *Radio New Zealand* in 2021 make the parameters of the food challenge in Aotearoa clear:

> People are going hungry even though New Zealand produces enough food to feed 40 million – and it's spurring calls for the country to "feed the five

> million first". Almost 40 per cent of New Zealand households experience food insecurity, while 19 per cent of children live in households that experience food insecurity.
>
> (Hancock, 2021)

The problem is familiar for anyone studying food insecurity in uniquely productive areas like the California Central Valley today (Fairbanks, 2021, p. 14) or indeed any student of the history of famines under imperial rule (Davis, 2002). Massive amounts of food are certainly being produced, increasingly on industrial-scale thousand-hectare horticulture operations or livestock ranches, and this food is first processed in an ever-smaller number of ever-larger national hubs, then sold to international markets that offer the most attractively profitable prices. "Could be, we'll have a North Island grower and a South Island grower, give it a few more years" one former horticulturalist remarked to me, regarding the sheer scale and level of monopolization in the industry. Another former grower noted that only crops sold on the export market receive government funding for research and development. Producing food for the local population is a secondary concern to selling to larger overseas markets.

The FIRE economy may have long eclipsed agricultural and industrial capital in Auckland, but across New Zealand commercial export agriculture remains an unusually big earner compared to food production in other "developed" nations. On the 80,000 hectares of horticulturally productive land, from the millions of beef and dairy cattle and tens of millions of lambs and sheep (Stats NZ, 2022), and by the work of more than 15,000 seasonal labourers brought in from the Pacific Islands and kept in on-site housing under temporary work schemes (Immigration New Zealand, 2022), the sector "creates" more than $3 billion NZD of GDP per financial quarter, around the same amount as the category of "financial and insurance services" and roughly two-thirds as much as the construction industry (Stats NZ, 2021). This is a major business for the country, and its prominence comes directly from selling to international buyers. That there may be hungry people who are much physically closer to the places this food is being grown, harvested, packed, and shipped is not a relevant concern for the producers if these residents cannot match the demand and the price that the export market offers. To attempt to make a more equitable food system in Aotearoa requires the same approach as has been previously discussed for housing affordability and urban sprawl: criteria other than maximal profitability must be demanded.

Spending on food is often cut back in the face of other, more inflexible demands on the family budget. One poverty researcher quoted by Hancock says of growing food insecurity in Auckland City:

> You've got families on minimum wage, who are paying market rent, who have to live in the city, because that is where their job is. Some of these people are essential workers, so you need them in a city if you want your city to function. And when the cost of living keeps rising but your wages don't, and when rent skyrockets, the food budget gets capped.
>
> (Hancock, 2021)

The story is similar in food insecure families in Melbourne (Kleve et al., 2018) and among the desperately poor households of agricultural workers in Fresno, California (Brown & Getz, 2011). If the choice is between losing access to shelter and transport or buying less food, the latter is understandably a common stopgap solution taken by many. In urban economies with deeply financialized housing stock and six-figure professional incomes, it is precisely the people commonly referred to as "essential workers" since the advent of the COVID-19 pandemic, those who do the systematically undervalued work that makes more abstracted professional lives possible, that suffer first. Instances of having inconsistent access to food is a common occurrence in New Zealand households, as Hancock noted in the opening of her article, and conditions for food security have become notably worse in Auckland in the past five years. As a result of rising living costs (again, housing is a significant contributor) and the instabilities created by the pandemic, the demand for food parcels from the charity Auckland City Mission has increased more than threefold in five years (see Figure 7.2), which in turn promoted other food charities to report that the demand for food parcels is dramatically outstripping supply (Radio New Zealand, 2022a). In Auckland, as in other regions that make and export far more food than is consumed locally, people are publicly wondering why profitable markets must necessarily be prioritized over desperate local need.

The impacts of such chronic and sustained food insecurity are well documented, in Auckland and in cities across the world. Susceptibility to disease, lifelong developmental disorders, and massive impacts on individual mental health and broader social outcomes are all linked to a household not having consistent access to a decent standard of food (Macaulay et al., 2022, p. 2). The people most drastically

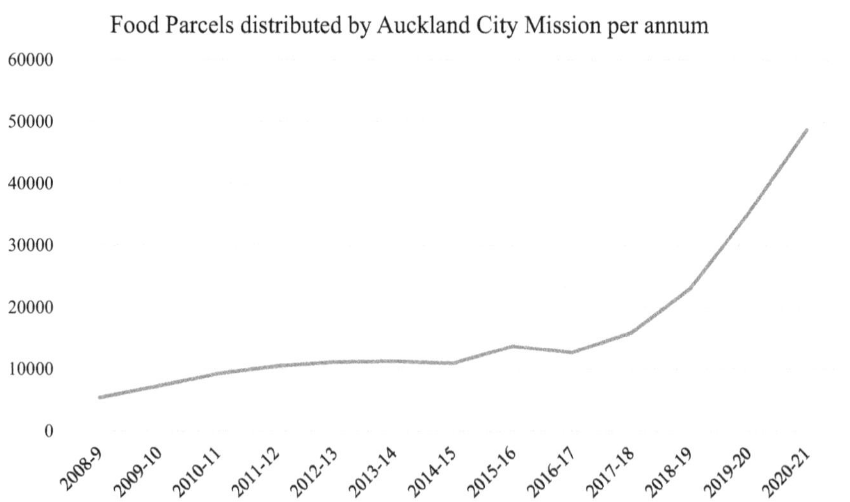


*Figure 7.2* Rising demand for food parcels from Auckland City Mission since 2008 (Child Poverty Action Group, 2021).

affected, here and overseas, are working single-parent families with children, most particularly those belonging to indigenous peoples and other marginalized ethnic groups (Robinson et al., 2021, p. 17). When the financialized periphery fragments eventually suburbanize horticultural land in Pukekohe, raising vegetable prices nationwide by 50 per cent as, indeed, we decide to import food rather than reduce sprawl, it is these most vulnerable residents of the fast-expanding city that suffer first and suffer the most. Beyond such notably vulnerable people living among us, however, the reality of daily food insecurity is not a hidden shame for an abjected few, but a simple fact of life for a large swathe of the people of Auckland. The difficulty of seeing in daily life the supposed third of households that face insecurity on a daily basis (Hancock, 2021) is partially due to hegemonic understandings of poverty in New Zealand and other "developed" nations. Dominant narratives (from individual agents and institutions intent on avoiding any redistributive politics taking shape in public discourse) present food poverty as a failure of budgeting, an inability for a parent to make intelligent choices, an individual weakness of character from parents who have personally failed their children, rather than a question of low wages, crumbling safety nets, and rampant inequality (Graham et al., 2018, p. 385). Those who cannot put sufficient healthy food on the table do their best to hide this fact from others to avoid such stigma, making it difficult to see food insecurity in public life and reinforcing the isolation and exclusion of being in this large and rapidly growing section of the working population (p. 386). When food production itself becomes defined by abstract financial logic rather than the needs of households and communities that make such profit-making possible in the first place, the business of agriculture stands in the way of ensuring people are fed.

While the ever-more concentrated, financialized, and monopolized food system that dominates horticulture in New Zealand is detrimental to the lives of its urban residents, small-scale horticulturalists are not exactly benefiting under this New Order either. Growers both current and former explained, as has been detailed in the first two chapters, that the success of massive horticulture operations in ideal growing areas far from urban infrastructure was built on the destruction or privatization of state-led co-operatives and the intentional loss of shared infrastructure support by these ventures. In the profit-first horticulture world, supermarkets and large exporters established a significantly more powerful position over producers and used this new leverage to favour the new, industrial-scale operations. Gesturing to a stack of crates packed with fruit, a grower complained of the crate suppliers and logistics companies "now they got preferred customers". They told me that the biggest growers now receive packing crates and other inputs rapidly and regularly, at the cost of taking the time to supply any of the smaller ones, "it used to be first come first served". Their co-worker, who had mostly preferred to let the other do the talking as I sat in their warehouse, simply said "the market, they prefer Hastings".[1]

This phenomenon is globally relevant. Without co-operatives or the alignment of state interests, individual food producers all over the world are easily outgunned by vertically integrated supermarket conglomerates with oligopolies over food distribution, marketing, and financing (Shaw & Shaw, 2019). To repeat a comment first bought up by a horticulturalist in Chapter 1, "you gotta make a deal with them ... they don't want you to die, of course. They keep your head above water, just". Profits for the supermarket are enabled by strict and often arbitrary control over what small producers' hand over, unless this oversight gets in the way of smooth operations for the supermarket. "One year, cabbage was short" a grower who works across Auckland tells me, before asking me what I think this means. I replied that the supermarkets still have to get it from somewhere, and they nodded. They detailed how a truck pulled up, took on their cabbages without inspection and faked certification, leaving them, as the grower, legally exposed if something were to happen with the quality of the produce. "I didn't have a deal with him [*driving the truck to the supermarket*], I had a deal with the broker, going over them like, you can't do that". Just as in any other form of capital accumulation, every barrier to a profitable and rapid cycle of returns – including the economic survival of small producers, the affordability of food for people that need it most, or the observation of food safety standards – must be transgressed.

The failure of Aotearoa's food production and distribution systems to actually provide food for people living in the country even as enough produce to feed the population tenfold is sold on export markets has led to a rise in public concern and calls for action. As well as the broad sentiment expressed in Hancock's piece that "feeding the five million" be prioritized over the profitable business of the agricultural sector, proposals for the regulation or even nationalization of New Zealand's supermarket giants have been entertained:

> The Government and New Zealanders have been very clear that the supermarket industry doesn't work. It's not competitive and shoppers aren't getting a fair deal. The duopoly needs to change, and we are preparing the necessary legislation to do that

Labour minister David Clark is quoted as saying in a government press release, promising to mandate pricing and introduce a new regulatory body for the sector (New Zealand Government, 2022). Speculation persisted for some time that a third supermarket brand such as the American Costco chain or Aldi from Germany, operating on a wholesale model, would be simply forcibly given access to the country's supermarket real estate in an attempt to create "real" capitalist competition and avoid the state having to appear as though it was directly splintering a duopoly (Radio New Zealand, 2022b). The labour government hopes their reforms will also end supermarket power over other, smaller, retailers, who have found that they increasingly needed to defer to the omnipresent supermarkets for access to contracts with food producers. As the food system stands today, however, the disempowerment of small food producers and the comparatively massive scale of the agriculture export market compared to any local demand makes it very easy for the food we eat to be as much a subject of financial imperatives as the forms of housing we live in and how we use our land. To find more equitable and less abstracted forms of food production, we must return to viewing cities themselves as crucial sites of horticulture.

## Rugose cities: the uniquely urban material basis of small-scale horticulture

A recurring set of oppositions threaded throughout discourses on urban sprawl in this book is one that contrasts rural production with urban speculation. Land that was once utilized to provide food to the residents of a large city is left fallow when it becomes more valuable as a speculative asset, horticulturalists complain that the expansion of suburban life into the periphery makes their work less viable, and professionals on all side of the debate over sprawl make dramatic predictions of an entirely urbanized future for places like Kumeu-Huapai, supported by distant and gigantic intensive food production operations. Yet, as has been mentioned in passing in my historical perspective on the Kumeu River Valley in Chapter 2, successful small-scale peri-urban horticulture is very much an urban phenomenon. This perspective does not deny the unmatched potential productivity of the rapidly fragmenting soils at Auckland's southern periphery that has dominated concern over sprawl in recent years, or indeed the unbeatable growing conditions of regions such as Hawke's Bay, Malborough, and Central Otago that have today been able to utilize access to national and international food markets in order to outcompete local growers. What it does propose is an understanding of how food can be – and has been in the past – grown to supply the population of a city under conditions that allow it to thrive on urban proximity, rather than being endlessly and irreversibly threatened by it. By returning to the case study of pre-1980s Kumeu-Huapai along with examples from Victoria and California, I will spend this section examining how a more equitable and sustainable food production system for urban populations than the current dominant form is not only possible but also existed in a limited capacity yet very visibly within living memory.

In 2018, Catherine Brinkley argued that the very shape that researchers and planners have tended to comprehend urban expansion as taking has drastically informed their understandings of sprawl and urban policy. For many sensible planners and geographers, if the fragmentation of land at the urban frontier is permanently destroying valuable ecosystem services and productive market gardens, an important step to mitigate the damage is reducing the surface area of the urban periphery. As has been described in Chapter 3, self-fulfilling speculative valuations of land are powerfully influenced by any adjacent parcels being successfully rezoned, a single "legitimated transgression" paving the way for others. As a result, urban policy on sprawl has notably been concerned with a "concentric" approach to zoning for expansion (Brinkley, 2018, p. 215) to minimize the amount of urbanizing land that is adjacent to the periphery. Rather than allowing deep incursions into rural land, suburban and then urban developments radiate out in a roughly circular form. This logic is very sound when it comes to the intractable conflict that planners and other professionals have fought every day for decades, that of an unending demand for land at the periphery of growing cities to be released for speculative suburban development. Yet, Brinkley pointed out, this defensive stance against voracious property development interests has led to the abandonment of some positive and irreplaceable aspects of what she calls the urban interface, the place where the city and the periphery engage with one another.

Pointing to the long-standing practices of green wedges in urban development, Brinkley presented the concept of "Rugosity" (literally meaning a very variable or "wrinkled" surface area) as an aspiration for a city and its foodbowl, rather than something to avoid. She made a direct comparison to rough, folded, high surface area interfaces found on absorbent structures in ecology and the medical sciences, seeing the urban interface "as an important metabolic barrier to be maximized" (p. 216). The unique advantages of maximum possible proximity to urban buyers, ease of access to equipment and other inputs, and the labour force and expertise of the city make small-scale horticulture at urban margins not only possible but also extremely productive. Studies on market garden operations around the world have noted far higher productive outputs per hectare than larger industrial growers can attest to (Haberman et al., 2014; McDougall et al., 2019), and as was also noted back in Chapter 2, in the late 1970s, the New Zealand geographer Warren Moran had found with some surprise that food producers only became more productive the closer they were to Auckland City proper, not less (Moran, 1980, p. 234). Far from the work of peri-urban horticulture being just an inevitable victim of its parochial inefficiency in the face of a monopolizing food system and endless property speculation, it offers forms of food production and economic interaction for the city that cannot be found anywhere else.

For small growers operating across the northwest of Auckland, success had always been directly associated with the city, even if its rapid financialized expansion was the second chief culprit for the end of their industry after the liberalization of the food system. If close proximity to rail lines and the infrastructure of the centrally managed Boards made it possible for the optimistically advertised Huapai Estate to actually become a productive powerhouse, then the spread of paved roads and the rising population of the northwest over the 1960s gave growers ever more access to city-dwellers looking for fresh produce. Through the selling of produce from the front of the farm to passers-by (Gate Sales, a common and unregulated activity) and the clandestine transfer of large amounts of produce to nearby grocers and markets without going through the Board (illegal and regularly clamped down on by agents of the Board), horticulturalists in the Kumeu River Valley and elsewhere on the periphery were not only making a living but also becoming keenly aware of the amount of money to be made from selling to urban markets and using urban labour and materials. The possibility of yet further profitability is of course what drove some of the northwest's more successful horticulturalists to support deregulation and privatization, as was described in Chapter 1.

The remaining operations in the northwest and elsewhere in Auckland today, when not struggling with inflexible and exploitative supermarket contracts, continue to seek out grocers, cafes, and other retailers as their preferred buyers. As noted earlier in this book, a local historian working on the history of growers in the Kumeu River Valley told me of the horticulturalist who had been juggling up to five other collapsed orchards at once "His last client was a Café", while the patriarch of one of the few extant operations is seeking to bring on young Deloitte-trained relatives, "a younger and more aggressive bunch", to sell their produce to a high-paying urban market by combining it with other high-value food items. Even as these growers continued to send in their produce for export through the Boards, they sought access to local urban markets as a central part of their business model, in contrast to the necessarily transnational and export-dominated perspective of larger industrial operations across New Zealand.

As with many others studying the productive urban periphery, Brinkley acknowledged that the existence of a productive "high rugosity" urban fringe depended on the ability to defend productive land from speculation and to provide horticulturalists the infrastructure to grow food for the city (Buxton & Butt, 2018; Curran-Cournane et al., 2018). However, in a rather astute if somewhat depressing observation, she pointed out that low rugosity, concentric cities in her study also faced a dramatic acceleration of peri-urban land and thus that the political economic challenge of restricting the financialization of the periphery cannot be sidestepped by keeping to concentric forms of urban expansion (2018, p. 223). Whether rezoning rural land for suburban and then urban development occurs in a broad circular wave of the least surface area or a series of fractal wedges and spikes with the most urban-rural interaction possible, our cities face the same seemingly intractable challenge. Brinkley also noted that several urban planning initiatives across North America and Europe have already internalized this and are acting on "the distinct and desirable partnerships between the two land-uses" (p. 223). From this viewpoint, the arguments made in the first half of the book – that financialized forms of development rendered places like the Kumeu River Valley non-productive in the service of providing maximally profitable forms of suburban development with minimum investment in necessary services and that this has only ever historically been averted with the restriction of speculative subdivisions and the alignment of state interests towards local production – offer the possibility not only of mitigating harm but also of providing an opportunity. If, given the forms of production that defined the mid-century interregnum and made peri-urban horticulture possible, the city can create a relationship with its foodbowl that is more symbiotic than self-destructively voracious, the food systems of growing cities across the planet could be reimagined.

What does a "high rugosity" city offer its residents when it comes to their access to food? For a concrete picture of the differences, our understandings of urban Food Security and Food Sovereignty (and their relevance to this situation) need to be discussed in more detail than was presented in this chapter's introduction. For people who live and work in cities, those "essential workers" who provide the necessary services to allow for the existence of financiers, real estate agents, property developers, and other agents of the FIRE economy in the first place, food security is a matter of access and availability (Birch & Keating, 2013, p. 7). Well before any considerations of the individuals "food culture" or "responsibility" can make an impact, any working resident of the city must be able to afford the food they need for themselves and their family to live a healthy life, and they have to actually be able to find it. Under a food system defined and directed by supermarket dominance after the collapse of local horticulture, both of these tasks become more difficult. As mentioned earlier in the chapter, smaller retailers and grocery stores in New Zealand are finding themselves having to turn to supermarkets to acquire fresh produce and thus finding themselves unable to compete with the duopoly, while the supermarket storefronts themselves become the only option for buying food for many urban residents. As a result, when the two supermarket conglomerates that control food provision in Aotearoa from greenhouse to pantry gain returns of 13 per cent per annum, with a combined "excess profit" (i.e. an amount of profit beyond the bounds of what is considered reasonable for a national supermarket chain to admit to) of some $430 million NZD every year during the same period of time when food parcel demand began to increase exponentially (Pullar-Strecker, 2022), no alternative can be proposed beyond promises of a new attempt at regulating the giants. Ideas around demanding a third supermarket, or a wholesale supermarket, only underline just how much the food insecurity of someone living in New Zealand, particularly one living in a city, is a product of the success of supermarkets and their preferred industrial-scale suppliers in redefining the national food system in their favour.

This situation – food insecurity among urban residents as a result of the collapse of local produce suppliers and the monopolization of the food system – is a global reality. In Fresno, that site of unmatched agricultural productivity along with unprecedented inequality and social segregation, researchers from various California universities have been noting the links between food poverty and the abstraction of the food system away from local production and consumption for decades (Brown & Getz, 2011; Fairbanks, 2021; Khachadourian, 1999). The fact that this insecurity is most desperate among agricultural workers (Minkoff-Zern, 2014; Wirth et al., 2007) is a particularly ironic example of the disconnection between producing massive quantities of food and actually being able to access and afford it. In the other case study area from Chapter 5, peri-urban Melbourne, food insecurity is a matter of compromising on the quality of food in the face of extremely high costs of living, creating and refining complex budgets to cut down on food to afford staying in a house (Kleve et al., 2018, p. 2214). In Japan, Yoshifumi Ikejima noted the prevalence of food deserts in the cities of wealthy "developed" nations such as Australia, the USA, and New Zealand, before examining food insecurity in the Hodogaya ward in Yokohama City. The results of their work led to them arguing directly and explicitly for the development of new sources of local urban and peri-urban food production to address a lack of access, hoping to use produce stands to fill the gaps supermarkets create (Ikejima, 2015). For all of these cases – cities in wealthy and agriculturally productive regions of "developed" nations where working people struggle to put healthy food on the table – food security is built on prioritizing the provision of food to residents from local sources over the abstracted financial goals of a supermarket conglomerate or exporter.

The latter group of the two, the ever-larger and ever-more-concentrated food producers who export several times more than what they sell on the domestic market, are particularly relevant to what a "high rugosity" city offers in terms of Food Sovereignty, rather than just Food security. Aotearoa exports significant quantities of fruit, vegetables, meat, and milk, while importing processed sugars and grains (Rush & Obolonkin, 2020). In other words, we (as with other foodbowls across the planet) are sending fresh food out of the country for maximum profitability, then buying up less healthy options from international suppliers to fill supermarket shelves, a major contributing factor to "why those who live in a land of plenty are not well nourished" (p. 308). At the same time, the supply of affordable fresh fruit and vegetables is genuinely threatened by the loss of the "last frontier" of peri-urban production at Auckland's southern border, as Deloitte (2019) warned and as has been discussed in Chapter 4. Food sovereignty is therefore important for Auckland residents in two different directions – a right for access to the large amount of food being exported out of the country and the urgent need for the region's food-producing land itself to remain productive. One fondly remembered aspect of life in the Kumeu River Valley decades past – even from interviewees who had welcomed the financialization of the periphery – was the ready availability of fresh produce of all kinds grown as a seasonal accompaniment to the years-long project of establishing an orchard. Interlocutors recalled buying potatoes, cabbages, stonefruit, apples, homemade wine and preserves, and a range of other local products, from places that to my own experience had long been fragmented, suburbanized, or even fully absorbed into the city. New suburbs like Riverhead, Whenuapai, and Hobsonville, industrial and commercial districts such as Lincoln and Rosebank Road, all within living memory produced food that made its way to local tables. In the much more abstracted food system we have since inherited, the connection between the city and its food sources is far more attenuated. As I noted all the way back in Chapter 1, by the time I was attending the Kumeu Show in early 2021, there remained a single fruit stand out of all the dozens of stalls and attractions, and this was an event hosted by area's Agricultural and Horticultural Society. The close interplay between residents and food growers established in a High Rugosity city, however, once provided access to, and indeed more of a direct control over, the produce that is actually being grown and sold in the land upon which we make our lives.

With this in mind, the seemingly unending battle to protect ecosystem services and productive land at the margins of growing cities is not merely, as one older resident had remarked to me, "King Canute holding back the Tide", a most-likely futile attempt at damage control. To attempt to curtail urban sprawl and provide affordable housing carries with it a complementary positive development, creating a more equitable and sustainable food system for everyone living and working across the city and its periphery. Yet, as has been noticeable for much of this book, these goals require actions that directly contradict the reigning economic interests of the state, the collection of individuals and interests often glossed as the "business community", and much of what makes the FIRE economy in Auckland possible. Even considering such potentially transformative policies for the food system and housing supply entails a confrontation with the increasingly disconnected and abstracted forms of economic activity that Gudeman (2016) termed the Financial and Meta-Financial spheres.

## Food in an unequal world: the financialized food system now and in the future

More than two decades ago, Keith Hart sought to explicate "the contradiction between agrarian civilization and the machine revolution of which capitalism is both the source and the outcome" (2000, p. 308). He believed that the people of the world, caught between the millennia-old agricultural institutions that continue to underpin human life and a rapidly evolving "machine capitalism" or "virtual capitalism", were being ever more deeply divided between a dwindling enclave of wealthy beneficiaries of financial extraction and a growing population of abjected working people struggling to survive in a crumbling world:

> a society so cruel and indifferent to the general human interest is heading for ecological disaster. Ours is a corrupt Ancien regime that must soon find a new democratic revolution, if human intervention in the life of this planet is not to end in catastrophe.
>
> (p. 310)

For Hart, one of the challenges facing humanity and the biosphere in the years to come was that the demands of the financial class and thus of global capitalism had grown beyond the ability of other interests to control. With the alignment of the powers of the state towards the protection of financial profit-making, barriers in the way of capital accumulation have been pulled apart at unprecedented rates since the restoration of financial power in the early 1980s, with dramatic results for human life and work. As Hart puts it, "This apotheosis of capital, its effective detachment from what real people do, has made many huge fortunes and produced devastating losses" (p. 313). At the frontier of growing cities across the planet, the conflict between the agrarian base of all urban life and the infinite demand of capitalists to find profitability in a finite world finds its clearest expression in the transformation of the urban food system.

When the chair of a pro-finance New Zealand think tank declares that we must logically build suburban housing on productive land and grow food in poorer soil, when a council member for a deeply segregated California city declares that the people have cried out for decades for market-rate single-family housing to be built in this city's poorest areas, and when Australian property developers create lifestyle block paradise in the path of seasonal fire corridors, who benefits and who suffers? The financialization of land (when it comes to both housing and food) has indeed "made many huge fortunes", and most of my interlocutors in Kumeu-Huapai, many of them older land-owning professionals, were recipients of some part of this wealth. The unmatched appreciation of the value of a house in Kumeu-Huapai between 2014 and 2019, amid a context of a general massive rise of house and land prices at Auckland's periphery in the past two decades, has made the financial success story of a liberalized economy a very concrete reality for these interviewees. Even for those horticulturalists who, as has been related in previous chapters, "went grumbling all the way to the bank" in the face of speculation and fragmentation, the compensation gained from being landowners at the periphery ensured they could become the owners of a new venture or simply retire early. The most ardent supporters of the financialized periphery, the owners of suburban development and housebuilding companies, saw the benefits of the urbanization process most clearly, and its costs in the most abstract terms. In response to a question about the profitability (for some) of urban sprawl into productive or protected areas, a manager of one firm told me over the phone in patient terms:

> because in New Zealand it's all free market … if the farmland becomes housing, then it becomes a higher demand, and so someone, like farmers, will enter the market to take that demand … it's a process of change, we are not locking down land.

For them, at least outwardly, the transformation of the periphery from food producer to investment vehicle is not only lucrative but also in harmony with a greater system of self-correcting development. That their success at the financial level would be related to a cost elsewhere is at best a hypothetical possibility, compared to very clear material evidence of profitability.

In contrast, the people who suffer the most from the dominance of finance-led property development and agriculture are those who have the least stake in its profitability, yet also provide the crucial labour and consumer demand that allow such structures to exist. This is an extremely broad collection of people and families, from the construction workers and farm labourers working as contractors with a fragile relationship to the accumulation of wealth that their employers are gaining from the land, to professionals finding themselves tied to a million-and-a-half dollar mortgage for a suburban home an hour's drive from their work and social life, to city residents losing the ability to keep up with living costs caught between rocketing property prices and an ever more abstracted and disconnected food system. Many people who live and work in urban areas across the world, particularly those who actually keep the city running on a daily basis, pay the price for sprawl.

It can be tempting to reify this conflict between the basis of human existence and capital accumulation into the imagery of forces, or waves or some other visual representation of monopoly and speculation. Yet here, in Kumeu-Huapai, as in countless other individual towns at the edge of growing cities across the planet, one can point to the interests of individual agents and of specific institutions working in the territory field of the urban fringe and draw a substantive picture of the collision that occupied Hart's image of the future at the turn of the millennium. The unprecedented enrichment of property holders and the success of the real estate sector have been built on converting peri-urban land into suburban house-and-land packages with the use of flexible contract labour, then selling them to those city-based professionals with the income to cover the terms of the mortgage. Every other concern – infrastructure, affordability, sustainability, and protection of Auckland's foodbowl – is a potential limiting factor on the spread of this model and thus has been overcome or sidestepped by the individuals seeking to make their fortune at the periphery, with the aid of state policy that protects financial outcomes. It is these "Ruling Interests, not Ruling Ideas" (Harvey, 2021) that engaged in the creative destruction (Harvey, 2007) of state-led peri-urban horticulture in order to allow for suburbanization and control over the food system. Even if, as one former horticulturalist had noted back in Chapter 1, some of these ambitious individuals had failed to secure the spoils of privatization, "ruining the Apple and Pear Board to build an empire that never got realized", those who did manage to take control of the situation drastically increased Aotearoa's potential for financial extraction (Kelsey, 2015, p. 47). This wealth was gained at the cost of the living conditions of the working people who had made the existence of such ambitions possible in the first place.

Having presented such a polarized and grim vision of urban food systems in Auckland and elsewhere, it is important to emphasize that due to the drastic nature of Aotearoa's own "mythical food crisis", critiques, analyses, and new possibilities abound. This is more true now at any other time in recent memory, as the myriad individual organizations that make up the global supply chain continue to shudder under the weight of the ongoing COVID-19 pandemic and imminent recession (Friesen, 2021). One of the local historians I had spoken to about the collapse of horticulture in the area made a point of contrasting the seemingly unstoppable process of privatization and globalization sweeping the industry in the 90s and the new millennium with the sudden cascade of failures and blockages threatening the export-led horticulture business. "I think we have seen that long global supply chains have an Achilles Heel … there are advantages in peri-urban [*horticulture*]". While the owners of large-scale horticulture operators demanded that immigrant workers be given expedited passage into the country (Radio New Zealand, 2021), the labour government tempered such moves with open concern about the extent to which the country's food system relied on flexible, cheap, and readily available international labour sources (Frykberg, 2021). With said labour less easily accessible, some large producers in New Zealand are being forced to return to the "family life cycle" model that O'Ryan (1992) noted once formed the centre of productive operations in Auckland's northwest, relying on the presence of younger relatives on the farm to get the same work done (Murphy, 2021). Other limiting factors, such as the cost of shipping (Thukral & Maguire, 2021) and indeed threats to local food security undermining the capacity of other nations to actually supply the people of Aotearoa with food (Gregory, 2009, p. 294; Rigg et al., 2020) bring more force and more attention to the arguments of those, like Dr Curran-Cournane (2018), who demand a reassessment of our relationship with land and with food. What may be often presented as intractable issues, simply inherent to the nature of cities, now require us to imagine urgent and transformational visions of the city and its periphery.

While everyone in Aotearoa must eat food to survive, the food crisis created by the financialization of the periphery and state abandonment of local production in favour of maximally profitable exports strikes the working poor the hardest. Increases in vegetable prices by 50 per cent from the development of Pukekohe may cause salaried professionals to hesitate at the counter, but for the large (and growing) portion of Auckland's population already struggling to put healthy food on the table and pay rent, the loss of the city's remaining peri-urban production will have a drastic effect on daily life. For the same reason, the forms of peri-urban production that Brinkley argued for, a food system based on a symbiotic relationship between an urban centre and its surrounds, offer forms of food security and food sovereignty that are most meaningful to those vulnerable urban and suburban residents who today must rely on a system much more interested in profitable exports than feeding nearby populations. The ability of some experts and development sector professionals to deny the existence or possibility of a food crisis in New Zealand rests with their (relative) insulation from this daily reality. At the heart of the conflict between peri-urban horticulture and speculative development, between Hart's "agricultural institutions" and "machine/virtual capitalism", is the desire of a group of professionals and capitalists to extract what they can from the periphery, at any cost. As the following chapter will show, this conflict at the urban periphery creates dramatic challenges to work and life in these formerly rural areas, not only due to the inequalities and planning failures created by financialization but also due to the increasingly unstable relationship our cities have with the land itself.

## Note

1 A city in the Hawke's Bay region, notable for horticulture and viticulture production.

## Bibliography


Birch, E. L., & Keating, A. (2013). *Feeding Cities: Food Security in a Rapidly Urbanizing World*. Penn Institute for Urban Research. Retrieved July 25 from https://penniur.upenn.edu/uploads/media/Feeding-Cities-Report.pdf

Brinkley, C. (2018). High rugosity cities: The geographic, economic and regulatory pathology of America's most non-concentric urban areas. *Land Use Policy*, *73*, 215–224.

Brown, S., & Getz, C. (2011). Farmworker food insecurity and the production of hunger in California. In A. H. Alkon & J. Agymen (Eds.), *Cultivating Food Justice: Race, Class, and Sustainability* (pp. 121–146). Cambridge: The MIT Press.

Buxton, M., & Butt, A. (2018). *The future of the fringe: The crisis in Peri-urban planning*. Melbourne: CSIRO Publishing.

Child Poverty Action Group. (2021). *Auckland City Mission Food Parcel Demand (2021)*. Retrieved July 27 from https://www.cpag.org.nz/statistics/auckland-city-mission-food-parcel-demand
Curran-Cournane, F., Golubiewski, N., & Buckthought, L. (2018). The odds appear stacked against versatile land: Can we change them? *New Zealand Journal of Agricultural Research 61*(3), 315–326.
Davis, M. (2002). *Late Victorian holocausts: El Niño famines and the making of the third world*. London and New York: Verso Books.
Deloitte. (2019). *New Zealand's Food Story: The Pukekohe Hub*. Retrieved from https://www2.deloitte.com/nz/en/pages/primary/articles/case-study-pukekohe-hub.html
Fairbanks, D. H. (2021). Geography of hunger: Demographic and socioeconomic determinates of large-scale severe hunger rates in Fresno, California. *Journal of Poverty*, *25*(1), 1–22.
Friesen, G. (2021). *No End in Sight for the COVID-Led Global Supply Chain Disruption*. Retrieved September 7 from https://www.forbes.com/sites/garthfriesen/2021/09/03/no-end-in-sight-for-the-covid-led-global-supply-chain-disruption/
Frykberg, E. (2021). *Govt to Allow 2000 Horticulture Workers in from Pacific under Strict Conditions-Radio New Zealand*. Retrieved August 9 from https://www.rnz.co.nz/news/national/431549/govt-to-allow-2000-horticulture-workers-in-from-pacific-under-strict-conditions
Graham, R., Hodgetts, D., Stolte, O., & Chamberlain, K. (2018). Hiding in plain sight: Experiences of food insecurity and rationing in New Zealand. *Food, Culture & Society*, *21*(3), 384–401.
Gregory, C. (2009). Whatever happened to economic anthropology? *The Australian Journal of Anthropology, 20*, 285–230.
Gudeman, S. (2016). *Anthropology and economy*. Cambridge University Press.
Haberman, D., Gillies, L., Canter, A., Rinner, V., Pancrazi, L., & Martellozzo, F. (2014). The potential of urban agriculture in Montréal: A quantitative assessment. *ISPRS International Journal of Geo-Information*, *3*(3), 1101–1117.
Hancock, F. (2021). *Calls to 'Feed the 5 Million First' before Exporting NZ Food-Radio New Zealand*. Retrieved July 27 from https://www.rnz.co.nz/news/whoseatingnewzealand/447324/calls-to-feed-the-5-million-first-before-exporting-nz-food
Hart, K. (2000). Money in an unequal world. *Anthropological Theory, 1*(3), 307–330.
Harvey, D. (2007). Neoliberalism as creative destruction. *The Annals of the American Academy of Political and Social Science*, *610*, 22–44.
Harvey, D. (2021). *The Anti-Capitalist Chronicles In [S3 E15] Dumping Surplus Capital*. Retrieved from https://www.youtube.com/watch?v=keD9m5EizsM
Ikejima, Y. (2015). The reality of food deserts in a large Japanese city and their resolution using urban agriculture. In S. Hongladarom (Ed.), *Food Security and Food Safety for the Twenty-First Century* (pp. 205–216). Singapore: Springer.
Immigration New Zealand. (2022). *Recognised Seasonal Employer (RSE) Scheme Research*. Retrieved July 27 from https://www.immigration.govt.nz/about-us/research-and-statistics/research-reports/recognised-seasonal-employer-rse-scheme
Kelsey, J. (2015). *The FIRE Economy: New Zealand's reckoning*. Wellington: Bridget Williams Books.
Khachadourian, I. R. (1999). *The prevalence of food insecurity among reentry women students and their children at California State University, Fresno*. Fresno: California State University.
Kleve, S., Booth, S., Z. E. D., & Palermo, C. (2018). Walking the food security tightrope—Exploring the experiences of low-to-middle income Melbourne households. *International Journal of Environmental Research and Public Health 15*(10), 2206–2225.

Macaulay, G. C., Simpson, J., Parnell, W., & Duncanson, M. (2022). Food insecurity as experienced by New Zealand women and their children. *Journal of the Royal Society of New Zealand*, 1–17. Retrieved from https://www.tandfonline.com/doi/full/10.1080/03036758.2022.2088574

McDougall, R., Kristiansen, P., & Rader, R. (2019). Small-scale urban agriculture results in high yields but requires judicious management of inputs to achieve sustainability. *Proceedings of the National Academy of Sciences*, *116*(1), 129–134.

Minkoff-Zern, L.-A. (2014). Hunger amidst plenty: Farmworker food insecurity and coping strategies in California. *Local Environment*, *19*(2), 204–219.

Moran, W. (1980). Mechanisms of land use allocation on the urban periphery, an Auckland case. In A. G. Anderson (Ed.), *The Land Our Future* (pp. 223–238). Auckland: Longman Paul.

Murphy, S. (2021). *Lack of Immigrant Workers Leaves Farmers to Rely on Family-Radio New Zealand.* Retrieved August 9 from https://www.rnz.co.nz/news/country/451919/lack-of-immigrant-workers-leaves-farmers-to-rely-on-family

New Zealand Government. (2022). *Government Acts on Supermarket Duopoly*. Retrieved July 29 from https://www.beehive.govt.nz/release/government-acts-supermarket-duopoly

O'Ryan, M. J. (1992). *Orcharding in the urban periphery: An examination of family orchard management strategies in Auckland's Northwestern periphery.* Auckland: The University of Auckland.

Partridge, R. (2019). *A Spoonful of Economics Helps Land Use Bias Go Down-National Business Review*, 27 September 2019. Retrieved June 8 from https://www.linkedin.com/pulse/spoonful-economics-helps-land-use-bias-go-down-review-roger-partridge?trk=d_public_post_promoted_post

Pullar-Strecker, T. (2022). *$430 Million a Year in Excess Profits But No Supermarket Split-Stuff NZ.* Retrieved August 4 from https://www.stuff.co.nz/business/127979707/430-million-a-year-in-excess-profits-but-no-supermarket-split

Radio New Zealand. (2021). *Plea for More Countries to Be Added to Visa Scheme for RSE Workers*. Retrieved August 9 from https://www.rnz.co.nz/news/country/451326/plea-for-more-countries-to-be-added-to-visa-scheme-for-rse-workers

Radio New Zealand. (2022a). *Food Charity Run by Farmers Says Demand Increasing Nationwide*. Retrieved July 27 from https://www.rnz.co.nz/news/country/470971/food-charity-run-by-farmers-says-demand-increasing-nationwide

Radio New Zealand. (2022b). *Supermarket Changes: Independent Wholesaler Needed, Say Retailers*. Retrieved July 29 from https://www.rnz.co.nz/news/national/468199/supermarket-changes-independent-wholesaler-needed-say-retailers

Rigg, J., Phongsiri, M., Promphakping, B., Salamanca, A., & Sripun, M. (2020). Who will tend the farm? Interrogating the ageing Asian farmer. *The Journal of Peasant Studies*, *47*(2), 306–325.

Robinson, H., Deane, K. L., Bartley, A., Alansari, M., & Neuwelt-Kearns, C. (2021). Shining a light on food insecurity in Aotearoa New Zealand: Modification of a food security scale for use with individuals who have extreme food security needs. *Aotearoa New Zealand Social Work*, *33*(4), 14–30.

Rush, E., & Obolonkin, V. (2020). Food exports and imports of New Zealand in relation to the food-based dietary guidelines. *European Journal of Clinical Nutrition*, *74*(2), 307–313.

Shaw, H. J., & Shaw, J. J. (2019). *Corporate social responsibility, social justice and the global food supply chain: Towards an ethical food policy for sustainable supermarkets*. London and New York: Routledge.

Stats NZ. (2021). *Gross Domestic Product: September 2021 Quarter*. Retrieved July 27 from https://www.stats.govt.nz/information-releases/gross-domestic-product-september-2021-quarter

Stats NZ. (2022). *NZ.Stat*. Retrieved May 18 from https://nzdotstat.stats.govt.nz/wbos/index.aspx

Thukral, N., & Maguire, G. (2021). *Double Whammy for Food Buyers as Freight Costs Spike Amid High Grain Prices-Reuters*. Retrieved August 9 from https://www.reuters.com/world/double-whammy-food-buyers-freight-costs-spike-amid-high-grain-prices-2021-07-09/

Wirth, C., Strochlic, R., & Getz, C. (2007). Hunger in the fields: Food insecurity among farmworkers in Fresno County. *California Institute for Rural Studies*, *13*, 1–30.

# 8 Climate change and the future of peri-urban development

In the early morning of the 31 August 2021, two weeks into a nationwide lockdown to halt the spread of COVID-19, nearly one-and-a-half times the entire monthly rainfall in the Kumeu River Valley fell in a few hours. The unusually heavy rain coincided with an especially high tide in the Kaipara Harbour, the body of water which the Kumeu River feeds into. With this tide raising the level of the river before the rain even fell, a once-in-a-generation flood was born. "It started to hit [*a few hours*] after midnight ... within a minute it was

*Figure 8.1* The author's brother rows past a neighbour's garage in the Ararimu Valley, captioned by the neighbour in a Snapchat message.

DOI: 10.4324/9781003400325-11

like a white-water rafting trip in here … stumps, tires, oil barrels" one resident recounted later, as they led me around a commercial lot on the main street where they worked and stored belongings. At the front door of the building, the dark stain of filth-ridden floodwaters sat some four feet off the ground. "I was in here for like three weeks drying out the walls". Sitting at home under the lockdown, my first knowledge of the flood was receiving a screenshot of a Snapchat message of my brother piloting a rowing boat past a neighbour's house in the nearby Ararimu Valley (see Figure 8.1). Buildings around the main street and the river-adjacent commercial areas were worst affected, but some residents in the town's southeast were surprised by the flood overtaking their properties from the opposite direction they expected and with unusual force. They pointed to the nearby rail line, which had been recently raised and reinforced, making a barrier for the flow of the water. "When they rebuilt the line last year they raised it … like a channel dam. It breached in four places". The lockdown wouldn't end until three months later, in early December, but even by this time, when I was finally able to return to Kumeu-Huapai, the houses of interlocutors still had sections cut out from walls where flood damage necessitated a replacement (see Figure 8.2), and piles of possessions that had been touched by flood waters and were therefore unsafe to keep. Many had to wait until the next year to receive their insurance.

*Figure 8.2* Floors and sections of the wall being cut out and replaced some four months after the flood. Photo taken by the author.

For the large majority of locals that have moved to Kumeu-Huapai in the past two decades, this sudden overnight deluge was the first real expression of the difficult relationship between the river and the lands of the valley, something otherwise only suggested in the calculations of property valuation experts and Auckland Council maps marking possible areas vulnerable to flooding (see Figure 8.3). For Derry Faulder, however, the flood was a familiar sight and a lifelong concern. As described back in the historical narrative of Chapter 2, Derry walked through the flooded main street as a teenager back in 1954, seeking shelter in the town hall. In the decades since, he took part in efforts to chart the extent of these generational flood events, heading out with several other locals on boats amidst another major flooding in the 1970s and marking out water levels on telephone poles. The plan at the time had been to use this survey to determine a 50-year-flood level for the area and limit development to this line. Yet, like the maps of productive land also made in the 1970s that Chapter 3 focussed on, these flood extent data were lost sometime between then and now, most likely in the reorganization of local councils under the single Auckland Council, according to Derry. Immediately after the flood, he told a reporter of the local newspaper:

> For over 50 years I've been a party, via several committees, to informing various district planners of historic flood levels … And as I sit in lockdown mode watching TV coverage of stricken families being bailed out of flooded

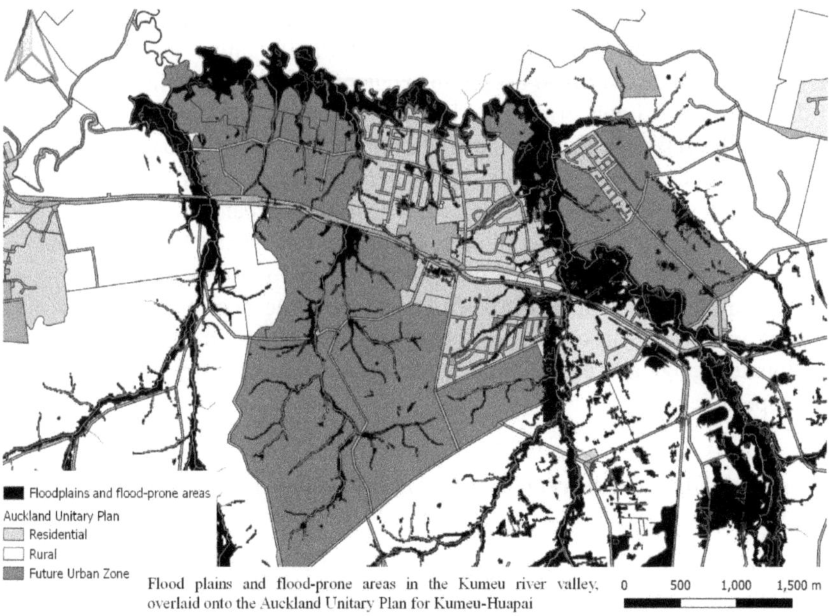


*Figure 8.3* Flood plains and flood-prone areas in the Kumeu River Valley charted by Auckland Council (Auckland Council, 2016) overlaid onto the Auckland Unitary Plan (Auckland Council, 2021).

> urban houses that have been given permits to build on formerly rural flood prone land, I feel the efforts have been in vain … I also note with apprehension last year's planning announcement of an urban village of potentially thousands of residential homes on 109 hectares of farmland zoned rural on both banks of the Kumeu River at Taupaki, on a known floodplain! Who would insure them?
>
> (Faulder, 2021)

From the point of view of one who has seen these flood events occur with some regularity several times over the course of their life, attempting to create a new suburb for Auckland in the Kumeu River Valley seems an exercise in hubris.

The business of suburban development in Kumeu-Huapai, like similar finance-drive projects across Auckland's new urban frontier and at the edge of cities in Victoria and California, is being driven into building houses on land that cannot safely accommodate its new residents. In their search for cheap land (Patel & Moore, 2017) at the urban periphery, developers are creating maximally profitable house-and-land packages in flood plains and fire corridors. Even without factoring in the effects of climate change, such practices endanger the lives of people living in the urban periphery in a far more immediate way than any reinforcement of class divisions, loss of food sovereignty, or lack of infrastructure can. When one does consider the destabilizing climate conditions of the approaching century, including the projected loss of a third of the planet's arable land by 2100 (Kummu et al., 2021) and the fact that sea levels are not only rising but the rate of their rise has doubled in the past half-century (Stats NZ, 2022), the decisions made over land uses at the edge of fast-growing cities become one of the most critical concerns faced by urban planners and policymakers today. This chapter presents a final perspective on financialized peri-urban development before the conclusion of the book. I consider the future of such a form of profit-making in a world where the accelerating pace of urbanization collides with an increasingly unpredictable climate. Returning once again to textual cases in Outer Melbourne and the California Central Valley to provide a basis for comparison and contrast with Auckland, I show how the dominant forms of new housing in these fast-growing peri-urban areas exacerbate the challenges posed by climate change and end by pointing to some alternative approaches to living on the periphery.

## Disaster in suburbia: extreme weather events in new peri-urban housing areas

> …Suburban firestorms are becoming more apocalyptic. The social cost of fire has increased in almost geometric relation to the linear growth of fire-suburb population. Two-thirds of all the homes and dwellings, for example, destroyed by wildfire since the state-wide record-keeping began in 1923 have been burnt since 1980.
>
> (Davis, 1995, p. 31)

Writing back in 1995 on the intrusion of upscale suburban living into Southern California's most fearsome fire corridors, Mike Davis believed the best solution was quite literally "Letting Malibu Burn". By returning to pre-colonial practices of regular small-scale burns, the exponential build-up in dry vegetation that encourages massive wildfires could be prevented from ever occurring. Davis noted, however, that attempts by the LA Fire Department and other authorities to put this policy in action were halted by a slew of legal actions from developers and landowners concerned about the impact regular burn-offs would have on their property values (p. 31). Today, in the new suburbs of California, as well as those in peri-urban Victoria and Auckland, the unending drive for maximally profitable development on land near growing cities is bringing this conflict – between ecosystem management and individual investment strategies – to the forefront of discussions around our relationship to the environment. Suburban housing developments established in flood plains and fire corridors are a calculated risk, factoring in the near certainty of a destructive event at some point in the foreseeable future with the opportunity for the conversion of rural land into extremely lucrative suburban housing. For land developers, housebuilders, financiers, and insurance agents, this risk is abstracted, a proposition of potential capital loss against the opportunity for further profit-making. The danger is far more material and urgent for the human beings who actually live and work in places like the Kumeu River Valley, those people who guarantee the profitability of suburban development by moving to these peri-urban frontiers and who ensure the functioning of the new settlement by providing services to the residents. Even when not considering the implications of climate change, the financial success of peri-urban development directly implicates people into situations where loss of life and property are an integral risk of the business.

Like the palatial residences of Hollywood's brightest stars in the second half of the 20th century, the new suburbs of California today have a close relationship with fire. Between 2000 and 2050, some 12 million acres of agricultural and park land will be broken up to meet the demand for new suburban and exurban development in the state of California (Mann et al., 2014, p. 438). In doing so, the large single-family homes of city-based professionals are becoming ever more directly exposed to local fire ecology. This loss of 12 million acres to development will not only destroy local horticulture and reinforce the financialization of land, as has been described in earlier chapters, but also place between 600,000 and 1.2 million new homes within "Very High Risk" wildfire areas (p. 446). The deeply unequal, extremely lucrative periphery of Fresno is among the areas of most dramatic and dangerous suburban growth in the state (p. 443). Life in the fire corridor does not simply threaten residents with the destruction of their home or threat to their life by means of a sudden inferno, it also inflicts long-term illness through the sheer quantity of smog produced by the burning scrub:

> The forest was burning, releasing a bounty of ash and energy the equivalent to 1,500 Hiroshima bombs into the Valley air basin. PM2.5 particles – small enough to bypass the lung's mucus membranes, enter the bloodstream

> and cross the blood-brain barrier – rose to a concentration that exceeded the EPA's annual limit by 600 per cent. The Valley, a headline in The [*Fresno*] Bee shouted, had the most polluted air in the world.
>
> (Weaver, 2021)

This segment from a local Fresno paper went on to explain that this air pollution caused permanent damage to the brain function, immune systems, and even DNA integrity of locals forced to breathe it in. Local government officials with interests in development, industrial agriculture, or oil extraction shout down attempts to present evidence of such externalities. One County Supervisor is reported yelling "I don't want to see your water, [*pounding his gavel*] I can take you to Riverdale and show you shit water, too!" when faced with a resident carrying a bottle of visibly polluted tap water from their house (*ibid.*). As Cade Cannedy put it in their thesis on air pollution in the area, the well-being of the people of Fresno is "undermined by capital-intensive production in order to increase private wealth" (Cannedy, 2020, p. 28).

In the outer reaches of Melbourne's exurban expanse, the ambitions of landowners and developers to sell regional Victoria to city-dwellers are running into similar barriers. In the two centuries of colonial Melbourne's existence, the regular appearance of massive and destructive wildfires has been a testament to the collision between, as Buxton et al. (2011, p.4) describe it, "[*an*] industrialised white settler society [and] the ecological reality of its natural context". The state of Victoria comprises 3 per cent of Australia's landmass, yet accounts for half of the nation's economic losses to wildfires (*ibid.*). As those seeking to make their fortune from the periphery clash with planners of the extending sprawl into wildfire areas, the "discretionary and permissive approach has led to a high degree of contestability in the decision-making process for land uses and developments and inconsistent decisions" (p. 6). The failure of governing institutions to restrict the ambitions of developers in Outer Melbourne has created a confrontation with Victoria's wildfire system, in much the same way that similarly flexible approaches to development enable the fragmentation of productive land.

For those living on the new urban frontiers of Auckland, the threat of a fast-moving inferno or a wall of toxic smoke is minor at the worst of times, at least compared to the situation in Victoria and California. A spokesperson for Ngāti Whātua o Kaipara's development trust noted with some concern the rise in small wildfires in the forests west of Kumeu-Huapai, "we've seen 145 fires out west last couple years", but the most immediate challenge by far comes from flooding. Two-thirds of the people across all of Aotearoa today live in a flood-prone area (NIWA, 2022). Many colonial settlements across the country were founded close to rivers for the obvious advantages lent by such proximity (McSaveney, 2017), only to, just like the settlers of Melbourne, collide with the ecological reality of the land they had taken. In places like Kumeu-Huapai, the felling of the rainforests for timber destabilized the relationship between the land and the water. From the first days of the Huapai estate and the attempts at horticulture in the river valley by the colonizers, the river broke its banks and threatened the new town with concerning frequency. "This is quite a regular thing, this" one resident remarked as we stood in a commercial area in between the main street and river, pointing to stormwater drains that would frequently back up and begin pumping rain runoff into the yard. At another time, a real estate agent expressed bemusement at the existence of the Kumeu Central development, sitting on the riverbank a kilometre or so east of the often-flooded commercial lot. "The water went everywhere it's not supposed to [*last August*] … they've always flooded down there on the new development". As the recent purchase by Fletcher of the former dairy farm south of Kumeu-Huapai on, as Derry described it, "both banks of the river" only further demonstrates, the possibility of unmatched profit-making (for the developer) on suburban development is overriding the material dangers (for the buyer) posed by these flood-prone areas.

Running through all three of these peri-urban cases is the common historical dynamic of finance-led development pushing past ecological limits for human habitation. On land expropriated from indigenous occupants and thus separated from their long-term understanding of the area's relationship to flood and fire, colonial speculators have spent centuries selling these unstable peripheries to city-dwellers looking for a new life in the countryside. In doing so, the risk is transferred from the developer to the new owners of suburban houses in fire corridors and flood plains. This dynamic has long defined development at the colonial urban periphery and has only intensified with the growing importance of real estate as an investment asset of last resort. Yet, this equation of risk and its accompanying practice of externalization, relying on insurance policy writers and mortgage brokers to take a chance on disaster, is being destabilized by the accelerating effects of climate change. Having shown the perilous business of peri-urban development in the past, the following section demonstrates that as extreme weather events become more commonplace and margins of safety are eroded, these suburbs at the edge of fast-growing cities face the full brunt of a much more hostile world.

## It gets worse: peri-urban life and food security in a destabilizing climate

Over the summer of 2022, the suburbs of the San Joaquin Valley cooked. Forty-one of 100 days in Fresno saw temperatures of above 100 degrees Fahrenheit (37.8°C) (Garibay & Ugwu-Oju, 2022). Nearly 90 per cent of the land area for Fresno County was classified as being in a state of "Exceptional Drought" (see Figure 8.4), where:

> Fields are left fallow; orchards are removed … fire season is very costly; number of fires and area burned are extensive … fish rescue and relocation begins … forest mortality is high; wetlands dry up; survival of native plants and animals is low.
>
> (National Integrated Drought Information System, 2022)

Amid this wide-scale destruction of food production, wildlife, and ecosystem services, the city's most vulnerable residents struggled to find shelter. Unable to

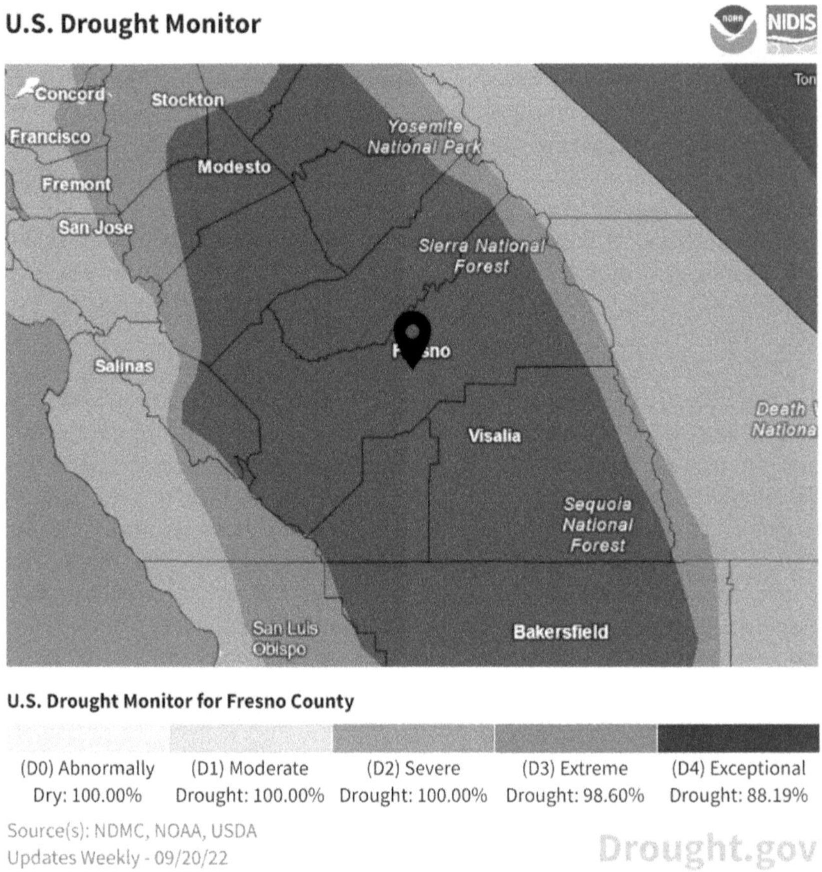


*Figure 8.4* Part of an NIDIS map of drought-affected areas in the California Central Valley in September 2022 (NIDIS, 2022). Note that the entirety of the valley is in a state of extreme drought, and most of it is under "Exceptional Drought" conditions. The U.S. Drought Monitor is jointly produced by the National Drought Mitigation Center at the University of Nebraska-Lincoln, the United States Department of Agriculture, and the National Oceanic and Atmospheric Administration. Map courtesy of NDMC.

afford a place with air conditioning, Fresno's poor working population went to local gyms and libraries, to canals, and to any shade they can find (Garibay & Ugwu-Oju, 2022). "With the heat, I have to go to hospital. If I start sweating a lot, I go to hospital … Sometimes it's too hot to rest (inside the apartment)" one resident told Cassandra Garibay, a reporter for the *Fresno Bee* (Garibay, 2022). The amount of acreage burned by wildfires has quadrupled in the past half-century, and as fires and the smoke they produce become a more common occurrence, the snows of the state's mountain peaks melt ever earlier, creating a rapid feedback loop beneath the burning of the land and the amount of heat absorbed from the sun

(Westerling, 2016). Even for those long-time residents who are used to the heat and smog of the valley, living in the summer is becoming harder.

As the effects of climate change begin to become more apparent in daily life, places that had once sat comfortably within the bounds of liveability are starting to teeter over the edge. In their efforts to forecast the effects of rising temperatures on land uses and food production, Kummu et al. (2021) presented the concept of a "Safe Climatic Space" to represent the parts of the world with "stable Holocene climate conditions" on which food production and human habitation rely. By their own calculations, a third of the land within this zone will become unsuitable for food production by the end of the 21st century, should carbon emissions continue on their current trajectory. This near-apocalyptic forecast of the world at the beginning of the 22nd century is relevant for the arguments of this book in two crucial ways: the importance of productive land close to cities and the instability of peri-urban life in a changing climate. Maximally profitable suburban sprawl, as this section will show, is taking away access to increasingly important sources of food security and sovereignty in the face of an approaching global food crisis, even as it places people and property in ever more danger of extreme weather events.

Speaking to a local real estate agent some months after the flood hit Kumeu-Huapai, I asked them for their predictions of the suburban housing market in the river valley in the coming decades. "Higher, in both respects. Prices will be higher and the houses are gonna have to be on stilts". A technical report commissioned by Auckland Council classified land in the path of the Kaipara River, into which the Kumeu River Valley's water flows, as having the highest possible classification of potential impact from climate change (Fernandez & Golubiewski, 2019, p. 18). Heavy rainfall in the area will assume a more tropical pattern in the coming decades, with dryer summers and less annual rainy days, but more sudden incidents of heavy rain during the wet season in the middle of the calendar year (NIWA, 2020, p. 94). With these facts known, and the severity of the issue reinforced by the recent floods, it is acknowledged that further suburban development should be accompanied by flood management infrastructure. One Council planner told me "The Kumeu Floodway Project is a prerequisite to [*further*] rezoning", referring to an attempt by Council to reshape the land around the river to protect against flooding, abandoned a decade ago due to complaints from local landowners (Morton, 2022). The Council's official report on the 2021 flood hopes that this work can be resumed and claims that "If completed, the floodway may have reduced flood depths by between 0.1 m and 0.86 m, depending on location" (Auckland Council, 2022). As these hypothetical improvements are discussed, development in Kumeu-Huapai continues and the Fletcher-owned former dairy farm splayed across the Kumeu River awaits a suburban future (see Figure 8.5). Peri-urban development in Kumeu-Huapai continues to grow at a geometric rate, even if some housebuilders may start to recommend to their clients that their new home be set up on stilts.

This pattern of development in the face of disaster is only a more dramatic example of how individual agents of the urban development sector in Auckland have confronted the externalities of their work. "Infrastructure comes last", a

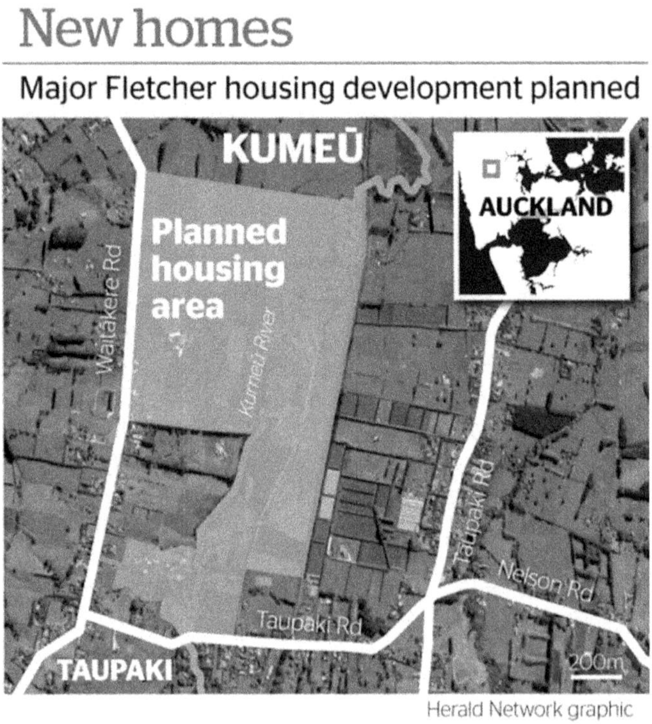


*Figure 8.5* Graphic from Ben Leahy's (2020) *New Zealand Herald* article on the Fletcher purchase of land directly in the Kumeu River for housing.

spokesperson for Ngāti Whātua o Kaipara's development trust remarked, whether the problem is traffic, jobs, housing affordability, forest fires, or floods. The "discretionary and permissive approach" to urbanization that has allowed the very profitable fragmentation of productive land at the urban periphery necessarily involves the developer's financial success being abstracted from the material results of their work. Looking at the growing pains of the rapidly expanding town and its uncertain relationship with the river, a real estate agent initially complained about the Council, pointing out they had a responsibility to provide support for private development, "they've done nothing … they put their heads in the sand". At the same time, they were distrustful of the developer Cabra's priorities for the area, "they don't have a great engagement or empathy with the community they work in … not a great deal of new parks going in" and were quick to disagree with any suggestion that the developer would hold back from further suburban development in river valley due to the issues created by their success. "There's [*still*] opportunities [*for*] countryside living around here". If further rezoning and expansion at Auckland's suburban frontier are legally and economically possible, landowners and developers will continue to push for it, even if they do so in the face of mounting costs for the people who come to live and work there.

In the new suburbs and lifestyle blocks of outer Melbourne, this dynamic is particularly clear when it comes to the increasing frequency of massive fires. Writing on the clash between the complex ecological system that defines the Victoria bush and the endless demand for land to be released for lifestyle development, Buxton, Goodman, and Moloney noted:

> The 2009 Victorian Bushfires Royal Commission understood the need for governments to anticipate future social consequences from past practice, stating that "to rebuild without any real thought being given to the future management of bushfire risk is to fail to learn from experience".

before pointing out "vacant rural lots in high fireprone areas are a time bomb awaiting the next conflagration, illustrating the need to severely limit development. Government support for continuing the incremental building of houses in such areas seems irresponsible" (2016, p. 93). Fire days in the region are estimated to increase by more than 40 per cent by 2050 (City of Melbourne, 2022), and in the face of this mounting threat, Buxton, Goodman, and Moloney argue that the fragmentation of rural land is the "key factor in this process" (2016, p. 91) of ever-greater exposure to risk. If the unmatched profitability of peri-urban sprawl encourages the subdivision and suburbanization of productive land and abstracts the very real risks of life in a fire corridor, then a crucial part of adaption to climate change will be confronting the social relations that make it both possible and profitable.

The urgency of this confrontation is only amplified by a growing concern from policymakers and academics about an oncoming global food crisis. While agricultural production becomes ever more concentrated and unequal (Reisman & Fairbairn, 2020, p. 690), and as arable land is lost to climate change in the decades ahead (Kummu et al., 2021), the world's population is forecast to continue to grow, reaching nine billion by the middle of the century (The Chicago Council on Global Affairs, 2013, p. 6). The people of the world will also be far more urbanized than they are even today. By 2030, some five billion of the world's eight billion inhabitants will live in cities (d'Amour et al., 2017). Amid predictions of a perfect storm of skyrocketing demand and crumbling supply across the planet, most notably in the parts of the world most vulnerable to the effects of climate change (Kummu et al., 2021, p. 7), feeding an urbanized population is perhaps the most immediate challenge facing humanity in this century. At the same time, reforming our global agricultural production system offers an important avenue for justice and emancipation, an opportunity to address planet-wide issues of land ownership equality, food sovereignty, and the labour rights of agricultural workers (Reisman & Fairbairn, 2020, p. 692). Just as the rising threat of wildfire and flood force us to understand how we interact with complex ecological systems, the promise of the destabilization of global food production requires a more equitable and less financialized approach to the urban periphery.

The unending demand for land at the edge of fast-growing cities to be released for suburban development has long brought house buyers into the path of disaster, but climate change promises to radically shift the calculations of risk in these unstable areas, even as it becomes more important than ever to preserve productive land. Finance-led processes of suburban development are particularly vulnerable to the threat of a destabilizing climate for the same reason they fail to provide the necessary infrastructure to support work and life: the pursuit of maximum profitability for landowners and developers over other material concerns. At the very same time as an accelerating pace of urbanization fragments formerly productive rural land, climate change also significantly reduces the amount of arable land on the planet, and the collision of these two processes promises to threaten the integrity of our extremely industrialized and concentrated global food production system. To mitigate the effects of this disruption and to adapt to a world with a climate far less congenial to human habitation, we must reassess how we use land on the urban periphery and how we make our lives there. As the following section will demonstrate, those who live in these vulnerable areas of the suburban frontier must already take steps to reorient their relationship to the land.

## Learning from a flood in the Kumeu River Valley

A few weeks before the August 31st flood, a representative of the local community action group admitted they had been worrying about how the changing climate would factor in life here. Bigger rainstorms (like the one that would strike soon after) could bring massive flooding without warning – "it's a flood plain all the way to helensville", while the growing frequency of wildfires in ever-drier summers kept firefighting crews on edge, "they go ballistic [*in summer*]". Speaking to them again three months later, in their home amid the disembowelled walls and the boxes full of household items that would have to be discarded, talk turned to the lessons learned and to preventative measures. First, in order to protect cash flow, insurance companies would always take as long as possible to respond to claims from suburban or lifestyle block homeowners struck by an unexpected deluge, "there's lots of loopholes for flood damage, water damage … they've [*the insurance companies*] all agreed, together, that they won't deal with it, especially in farms but also in residential areas". Second, the built environment of housing in the river valley would have to change, with elevated electronics such as air conditioning and pumps, supported by a lower structure made entirely of solid wood rather than composites. Third, if people intended to keep living here, especially in the low-lying eastern half of Kumeu-Huapai, then there needed to be a way to know what the river was doing and how to respond to it. The idea, explained to me by the community action representative and then repeated publicly on *Radio New Zealand* by Local Board member Phelan Pirrie (Radio New Zealand, 2022), is to set up a series of individual sensors on posts across the river valley, all connected by a "Low Power Wide Area Network" system, to collectively gather water level data during heavy rainfall events. They estimate these posts would cost about $150 NZD each, and "the more you sell [to locals], the more accurate it would be". If the development will continue apace even as flood risks worsen and infrastructure fails to materialize, then this locally led solution aiming to simply gain a few minutes warning is the most immediate adaption that residents can incorporate into their lives.

Measuring water levels and sending out a community alert in the event of a sudden rise can certainly provide valuable time for residents in the path of another flood, but this warning system does not reduce the severity of the other two lessons learned in the aftermath of the August 31st flood: the response of the insurance sector and the failures of the built environment. The former was, according to interlocutors who had suffered flood damage to their property, the most demoralizing aspect of the entire disaster. As possessions were thrown out and black mould began to grow in the walls of flood-damaged houses, residents found that every individual step in receiving their insurance payout took weeks. "Waiting, waiting, two weeks, then waiting", one lifestyle block resident recalled, before noting their elderly mother had been displaced by the flood and then rendered ill by the post-flood conditions, "it very nearly killed her, it was this close". The community action representative believed that the stress caused by the flood was far outweighed by that of dealing with insurers. "The trauma comes from the insurance company, if this was all handled quickly, I'd actually be feeling quite secure", another resident simply said, "it was very stressful … you start crying your eyes out while driving". In order to proceed claims, residents called their insurer nearly every day from the time of the floods until early 2022, while insurance agents did their best to protect their cash flow by asking for further clarifications, moving cases between agents, and seeking second opinions on site reports. The destabilizing climate is already leading NZ insurers to consider a "retreat" from covering some parts of the country (Cardwell, 2022), and as Derry had pointed out in Kumeu-Huapai's local paper, who would insure (at any affordable rate) houses knowingly built in a floodplain in the face of the promise of ever heavier winter rainfall? As the most recent floods have demonstrated, relying on the insurance sector to pay out at an increasing frequency leads the insurer to forestall payment as long as possible, to the detriment of the people actually affected by these extreme weather events.

The other costly lesson of the flood, that of the relationship between the built environment and the water, has more complex implications than local preparedness and nervous insurers. Many of the most affected residential properties were the southeastern extent of Kumeu-Huapai, low, flat land in the path of the river. The houses built on the lifestyle blocks here are dramatic examples of the risk calculations taken by peri-urban development even before climate change makes its presence known: flooding occurs regularly across the fields and villas, and most of the landowners in the area are, as one interlocutor pointed out in the previous chapter, absentee land bankers awaiting a profitable sale when the town's growth reaches them. The older dwellings (generally between 30 and 100 years old) that sit on these plots are built under far less rigorous standards than are enforced in such areas today. In response to my question about how the Huapai Triangle fared under the flood, one of Cabra's directors remarked, in "It all those older places that are worse off … everything that's new is designed accordingly". The existence of the other major development, Kumeu Centre, so close to the banks of the river, on land "that has always flooded" does contradict this, but it remains a fact that the low-lying lifestyle blocks around Boord crescent were badly affected by the redirection of the flood towards their land by the raising of the railway, in a way not seen in any other place that had been deemed fit for residential development. As conditions become more unpredictable, early victims of disaster will be not only the new developments that have most hubristically been built in the path of the flood but also the residences subdivided and sold on the promise of a manageable risk that now find the calculations running against their favour.

Even as these points have been driven home in the most direct possible way for so many of Kumeu-Huapai's residents, the endless demand for the subdivision and development of land remains a crucial part of the town's future for interlocutors. In order to reconcile these conflicting pictures of the river valley – increasingly unstable flood plain and new suburb of Auckland – developers and landowners emphasize the importance of state investment in flood protection systems for the growing town. In the historical perspective presented back in Chapter 2, profitable development in the river valley has repeatedly had to overcome ecological and social barriers to accumulation, and I ended that chapter by noting how private interests now sought aid from what remained of the state budget to provide the necessary infrastructure for further suburbanization. This attitude is also applicable to their response to flooding. As well as the council planner describing the abandoned local floodway project as a "prerequisite to [*further*] rezoning" a Cabra director expressed concern for the town's water infrastructure in future, saying Kumeu-Huapai needed "some more stormwater work", while a local official reported to a community meeting that Fletcher had told them the company intended to "sit on" the land they purchased in the Taupaki flood plain, waiting to see what happens with further development. Like the public infrastructure that had been promised to accompany Kumeu-Huapai's expansion but never materialized, these expressions of support for flood protection measures are (in the opinion of the author and a few interlocutors) efforts to allow further "legitimated transgressions" of rezoning rather than commitments to spending resources on creating flood management systems. Finding success with maximally profitable suburban development in a flood plain requires that developers separate their own work from the risk of flooding, just as have done so with traffic and other externalities of suburbanization.

In the aftermath of the August 31st flood, residents of Kumeu-Huapai confronted the unavoidable reality of the river valley's generational floods and the knowledge that such events would only increase in frequency. While some locals and officials work on an ad-hoc early warning system to give people minutes of valuable warning time, the failures of the nation's insurance sector to step in after the flood damage and the vulnerability of many lifestyle block dwellings to the deluge are creating concern about the future of peri-urban life in a destabilizing climate. Such pressures do not change the calculus of profitability for landowners and the development sector itself, however, and individuals involved in the business of property development at Auckland's urban periphery continue to seek what opportunities they can from the Kumeu River Valley. Adapting to a world with a less hospitable climate and a smaller area of arable land requires a reassessment of how to value and use the land that sits at the edge of our growing cities. Therefore, the final section of this last chapter before the conclusion looks at the options the urban periphery has in the face of a much more unpredictable biosphere.

**Future possibilities in the face of fire, flood, and drought**

In the same article where she investigated how California's ever-stronger heatwaves impacted Fresno's most vulnerable residents, reporter Cassandra Garibay noted the advice of a climate researcher on the situation: "While we don't have a lot of control over the changing climate at the moment, what we do have control over is the hyper-local area of how you can mitigate it and how you could adapt" (Garibay, 2022). In Fresno, the physical structure of the suburban landscape is a crucial factor in why its residents and working population are finding it so difficult to cope with these terrible heatwaves. The privatized suburban sprawl, the transport difficulties, and the concentration of wealth in professional-managerial retreats are all factors not only towards the destruction of local horticulture and the reinforcement of class lines but of vulnerability to climate change. As previous chapters have noted when addressing food security, urban form, or housing affordability in growing cities, confronting the issues created by the dominant forms of peri-urban development also offers an opportunity to improve life and work for our growing urban population at a time when conditions promise to become less favourable. For places like suburban Fresno, the Macedon Ranges, and the Kumeu River Valley, this challenge – and the concomitant possibility of drastically different futures for their residents – is ever harder to ignore as climate change becomes more visible. In the rapidly fragmenting peripheries of the planet's growing cities, choices made over land use now will determine the shape of life for urban residents of the century to come.

"This always was a shapeless town, a service town without a heart" the community action representative said, as they wondered at Kumeu-Huapai's uneasy relationship with the river. They noted how both lifestyle blocks and suburban developments had been "imposed on the township" in such a way that they faced away from the river, "driving in, you wouldn't know where it [*the river*] runs". One hope they held towards future construction was that it would be actually oriented around the Kumeu River, rather than ignoring it. This idea was also expressed in remarkably similar terms by the Ngāti Whātua o Kaipara development trust's environmental spokesperson, this time in regard to the northwestern highway around which Kumeu-Huapai is built. "It's how about you make sense of a place" they told me, pointing out that the highway roughly follows the path from the Kaipara coast to Auckland that once made the Kumeu River Valley an important portage, and yet the actual path of the highway (that also comprises Kumeu-Huapai's main street) cuts across the river and obscures its course from those who pass through here. The spokesperson believed that an important part of reassessing the town's relationship with the river would be rectifying this. In the opinion of both of these interlocutors, this reorientation of the built environment towards its ecological context is a symbolic improvement – an attempt to connect to the landscape which, as noted back in Chapter 3, has been repeatedly remade and in turn abstracted and obscured by new "structured permanences" (Mitchell, 2003, p. 16) constructed to sell the river valley. At the same time, however, this demands material changes to the relationship that residents and developers have with Kumeu-Huapai's land. Rather than using the river as a stormwater drain and a dump while building suburban housing across its banks, hoping that a massive flood will not strike in the selling period for the project, placing the river within the context of the growing town requires that industry and development operate under the constraints that the river valley places on us. Derry and his compatriots sought to ensure this half a century ago, when they measured water levels in the aftermath of another terrible flood, and proposed limits on development based on these markers. If Kumeu-Huapai, whether as a rural town or a suburb of Auckland, is to find a shape, to orient itself around a "heart", the river that made the town and threatens its expansion is the place to start.

This challenge of recognizing the ecological context that undergirds peri-urban land, understanding the agency of fire and flood, is part of what drove Mike Davis to demand that we "let Malibu burn" and encouraged the Victorian Bushfires Royal Commission to label Melbourne's state-supported sprawl "irresponsible". Many individuals have made their fortunes from transforming the land of the urban periphery into maximally profitable forms of suburban and rural lifestyle living, but as this book has repeatedly asserted, these profits are made possible by externalizing the costs of such forms of development onto the people who will work and live in these areas in the future. As drought, flood, fire, and other extreme weather events become harder to avoid in the new suburbs of growing cities, and as matters of urban food security assume an urgent tone for an increasingly urbanized global population, this contradiction is much harder to maintain. If insurers are considering a tactical retreat from vulnerable peri-urban developments, then the entire system of financial relationships that make building suburban houses on rural land in these places promises to be jeopardized as conditions become less predictable. Planners and policymakers can meet this oncoming challenge today by refusing to abstract suburban sprawl from the collapse of local horticulture and the cost of natural disasters.

In fact, a new approach to land use on the urban periphery is an immediate priority for any planners and policymakers seeking to adapt their cities to a changing climate. Over the course of this book, I have presented the critiques and solutions of those who seek a reassessment of how we treat the land at the edge of our cities. From Curren-Cournane's modified "farm-scale" approach to Land Use Capability and Brinkley's concept of Rugose Cities, to Murphy's analysis of the close relationship between development interest and government, and Fernandez's arguments against land supply as the key housing affordability issue, to the complaints, experiences, and theories of Kumeu-Huapai's residents amid an unprecedented transformation of the place where they live and work, these points have circled around this single conflict: peri-urban development is massively, unbeatably profitable for a small number of landowners and developers, and it offers professional-managerial families the chance to bet on their family home, while being extremely destructive to the material foundations of urban life. It is a visible and dramatic process that is contributing to the loss of productive land close to cities, to unaffordable urban housing, and to widening inequality. As Kumeu-Huapai grows 15-fold by the middle of the century (Rodney Local Board, 2017, p. 7), the town (or suburb, depending on how you want to define it) will run headlong into the shrinking ecological limits and infrastructure capacity of the river valley, as will so many other peri-urban settlements across the planet. It is here, on this land between the city and the country, where a more sustainable and more survivable future can be found.

When a flood struck Kumeu-Huapai in the early hours of 31 August 2021, a new generation of residents were forcibly reminded of the Kumeu River Valley's long history of flooding. Like other colonial peri-urban settlements across New Zealand and in places like Victoria, Australia, and California, USA, Kumeu-Huapai has had a difficult relationship with its ecological context since it was settled by Europeans. For decades, these places at the margins of growing cities have been a useful source of "cheap" rural land on which to build maximally profitable suburban developments and then leave the risks to the new residents and the insurers. As our climate starts to destabilize, however, this calculated move on the behalf of developers and landowners becomes more dangerous for the people who actually live and work in the settlements that they build and sell. Alongside forecasts of drastic losses to the planet's arable land in coming decades, this threat requires us to see the suburbanization of peri-urban land as one of the most immediate challenges in humanity's collective struggle to adapt to a world with less favourable living conditions.

## Bibliography

Auckland Council. (2016). *Auckland Council Geomaps*. Retrieved from https://geomapspublic.aucklandcouncil.govt.nz/viewer/index.html

Auckland Council. (2021). *Auckland Unitary Plan Operative in Part*. Retrieved September 10 from https://unitaryplan.aucklandcouncil.govt.nz/pages/plan/Book.aspx?exhibit=AucklandUnitaryPlan_Print

Auckland Council. (2022). *Kumeū Flooding Review*. Retrieved October 3 from https://www.aucklandcouncil.govt.nz/plans-projects-policies-reports-bylaws/our-projects/projects-west-auckland/Pages/kumeu-flooding-review.aspx

Buxton, M., Goodman, R., & Moloney, S. (2016). *Planning Melbourne: Lessons for a sustainable city*. Melbourne: CSIRO Publishing.

Buxton, M., Haynes, R., Mercer, D., & Butt, A. (2011). Vulnerability to bushfire risk at Melbourne's Urban Fringe: The failure of regulatory kand use planning. *Geographical Research 49*(2), 1–12.

Cannedy, C. (2020). *Why Your Air Will Not Get Better: Path Dependence and Capture in Air Quality Regulation Stanford*. Retrieved from https://west.stanford.edu/works/why-your-air-will-not-get-better-path-dependence-and-capture-air-quality-regulation

Cardwell, H. (2022). *Insurance Costs Could Increase Massively as Climate Risk Rises – Modeller – Radio New Zealand*. Retrieved October 25 from https://www.rnz.co.nz/news/national/473005/insurance-costs-could-increase-massively-as-climate-risk-rises-modeller

City of Melbourne. (2022). *Climate Change Impacts on Melbourne*. Retrieved October 4 from https://www.melbourne.vic.gov.au/about-council/vision-goals/eco-city/Pages/adapting-to-climate-change.aspx

d'Amour, C. B., Reitsma, F., Baiocchi, G., Barthel, S., Güneralp, B., Erb, K.-H., Haberl, H., Creutzig, F., & Seto, K. C. (2017). Future urban land expansion and implications for global croplands. *PNAS, 114*(34), 8939–8944.

Davis, M. (1995). The case for letting Malibu burn. *Environmental History Review*, *19*(2), 1–36.

Faulder, D. (2021). *Flood of Opinion – Kumeu Courier*. Retrieved October 28 from https://www.kumeucourier.co.nz/uploads/issues/october-2021.pdf

Fernandez, M. A., & Golubiewski, N. (2019). *An assessment of vulnerability to climate change in Auckland*. Auckland Council, Te Kaunihera o Tāmaki Makaurau.

Garibay, C. (2022). *Valley Towns Are Going to Stay Hotter for Longer. Here's What That Means for Your Health – Fresno Bee*. Retrieved September 30 from https://www.fresnobee.com/fresnoland/article264860474.html

Garibay, C., & Ugwu-Oju, D. (2022). *San Joaquin Valley's Most Vulnerable Communities Are Hit Hardest by Climate Change – Fresno Bee*. Retrieved September 30 from https://www.fresnobee.com/fresnoland/article264655664.html

Kummu, M., Heino, M., Taka, M., Varis, O., & Viviroli, D. (2021). Climate change risks pushing one-third of global food production outside the safe climatic space. *One Earth*, *4*(5), 720–729.

Leahy, B. (2020). *Fletcher Plans Huge New Housing Development in Kumeu, West Auckland, but Residents Up in Arms*. Retrieved October 3 from https://www.nzherald.co.nz/nz/fletcher-plans-huge-new-housing-development-in-kumeu-west-auckland-but-residents-up-in-arms/FROJPAX3BDOWYRZ2BH5QQICHY4/

Mann, M. L., Berck, P., Moritz, M. A., Batllori, E., Baldwin, J. G., Gately, C. K., & Cameron, D. R. (2014). Modeling residential development in California from 2000 to 2050: Integrating wildfire risk, wildland and agricultural encroachment. *Land Use Policy*, *41*, 438–452.

McSaveney, E. (2017). *'Floods – Flood control', Te Ara – The Encyclopedia of New Zealand*. http://www.TeAra.govt.nz/en/floods/page-6

Mitchell, D. (2003). Dead labor and the political economy of landscape – California living, California dying. In M. Domosh, S. Pile, & N. Thrift (Eds.), *Handbook of cultural geography* (pp. 233–248). Thousand Oaks, CA: Sage Publications.

Morton, N. (2022). *Kumeū Floodway Work Could Resume, a Year on from Devastating Storm*. Retrieved October 3 from https://www.stuff.co.nz/environment/300670626/kume-floodway-work-could-resume-a-year-on-from-devastating-storm

National Integrated Drought Information System. (2022). *Drought Conditions for Fresno County*. Retrieved from https://www.drought.gov/states/California/county/Fresno

NIDIS. (2022). *Current U.S. Drought Monitor Conditions for California*. Retrieved 29 September from https://www.drought.gov/states/california

NIWA. (2020). *Auckland Region Climate Change Projections and Impacts. Prepared for Auckland Council, Council-Controlled Organisations, and District Health Boards*. Revised September 2020. Retrieved October 3 from https://knowledgeauckland.org.nz/publications/auckland-region-climate-change-projections-and-impacts-revised-september-2020/

NIWA. (2022). *A Hard Rain's A'Gonna Fall*. Retrieved from https://niwa.co.nz/news/a-hard-rains-agonna-fall

Patel, R., & Moore, J. W. (2017). *A history of the world in seven cheap things: A guide to capitalism, nature, and the future of the planet*. University of California Press.

Radio New Zealand. (2022). *Renewed Calls for a Flood Warning System in West Auckland – Afternoons with Jesse Mulligan*. Retrieved October 28 from https://www.rnz.co.nz/national/programmes/afternoons/audio/2018835366/renewed-calls-for-a-flood-warning-system-in-west-auckland

Reisman, E., & Fairbairn, M. (2020). Agri-food systems and the Anthropocene. *Annals of the American Association of Geographers*, *111*(3), 687–697.

Rodney Local Board. (2017). *Kumeu-Huapai Centre Plan*. A. Council. Retrieved from https://www.aucklandcouncil.govt.nz/plans-projects-policies-reports-bylaws/our-plans-strategies/place-based-plans/Documents/kumeu-huapai-centre-plan.pdf

Stats NZ. (2022). *Coastal Sea-Level Rise*. Retrieved 23 September from https://www.stats.govt.nz/indicators/coastal-sea-level-rise

Emerging Leaders Class of 2013 (2013). *Feeding an Urban World: A Call to Action*. Chicago, IL: The Chicago Council on Global Affairs. Retrieved 31 March 2021 from https://www.researchgate.net/publication/335149467_Feeding_an_Urban_World_A_Call_To_Action

Weaver, G. (2021). *The Fallout Zone: Playing Autumn Sports in the San Joaquin Valley's Dangerous Air – Fresno Bee*. Retrieved from https://www.fresnobee.com/opinion/readers-opinion/article255943552.html

Westerling, A. L. (2016). Increasing western US forest wildfire activity: Sensitivity to changes in the timing of spring. *Philosophical Transactions of the Royal Society B: Biological Sciences*, *371*(1696), 20150178.

# 9 Conclusion

## Cities on the edge

The unmatched profitability of suburban development on rural land relies on a number of conditions and arrangements to ensure its existence. In growing urban centres all over the planet, this business model is sustained by flexible zoning laws, close relationships between the development sector and governance, the interests of financiers and landowners, and, perhaps most glaringly, the separation of the industry's profits from the costs it inflicts on the people and the land. The accelerating pace of peri-urban sprawl has made the urban periphery one of the most important sites of conflict over the future of human life, but not because it was an inevitable, natural result of an urbanizing population. Instead, the process that transformed places like Kumeu-Huapai is driven by the interests of individuals and organizations seeking to benefit from the creative destruction of productive land at the urban fringe and the unending demand for housing as a financial asset. Building a more resilient and equitable urban environment in the face of a destabilizing planetary climate and an increasingly abstracted global economic system requires an understanding of how this form of urbanization exists for the benefit of certain parties, and how these social and historical factors can be changed.

This book has argued that maximally profitable forms of suburban development draw their success from the intentional abstraction of their forms of production, circulation, and consumption away from their impact on the foundations of human existence. I began by presenting the frantic suburban and lifestyle development at Auckland's northwestern periphery alongside the collapse of local horticulture in the same area to show the acceleration of peri-urban development is an explicitly financial project. From this starting point, I placed the recent acceleration of financialized development in context with the colonial history of the Kumeu River Valley, showing that land banking and speculation have transformed its landscape repeatedly over the past 150 years and that Kumeu-Huapai's success in horticultural production relied on urban markets and the alignment of state interests and was likewise undone by them. I then detailed the efforts by developers and landowners to reshape the axiologies of peri-urban land in order to encourage rural fragmentation and allow suburban development, before pointing to more promising models of land use capability and legislation that could prevent this vicious cycle of speculation and productive collapse. Part One of this book ended with an analysis of the political economy of Auckland's urban development as it relates

DOI: 10.4324/9781003400325-12

to Kumeu-Huapai, noting how attempts to mitigate New Zealand's housing crisis for more than a decade failed to confront the financial drivers of housing unaffordability due to the dominance of the FIRE economy – Finance, Insurance, Real Estate – over life in Auckland City.

Having used the case of Kumeu-Huapai to look at the dynamics of peri-urban development in formerly productive rural areas, Part Two of this book brought in supplementary case studies from elsewhere in Aotearoa and from Victoria, Australia, and California, USA. Starting with the regional and national effects of Auckland's financialized urban expansion, I then used the cascade of issues that plagued the development reality show *The Block* in Gisborne, north of Melbourne, to show how the successes of suburban sprawl create barriers to further profitable accumulation. From the other international case in Fresno California, I demonstrated that the profitability of suburban development for financiers and landowners relies on the concentration of its costs onto the heads of the area's most vulnerable residents. From here, I delved into how peri-urban development drives the class divisions that were so apparent earlier in the Fresno case, pinpointing the professional-managerial class as the crucial source of effective demand that allows such impressive profitability and describing how these new homeowners seek to protect their property values by enforcing social and spatial barriers to non-homeowners taking part in community life in the new suburbs of the urban frontier. These cases continued to be used to examine how the destruction of local horticulture and the resulting monopolization of the food production system leave working people starving in a land that makes ten times the food it needs. I argued that a solution lies in seeing successful small-scale horticulture as a uniquely urban phenomenon. The final main chapter of the book addressed the extremely unstable relationship that these suburban developments have with their own ecological context. I used the most recent in a series of massive floods in the Kumeu River Valley alongside the mounting threat of wildfires in suburban California and Victoria to argue that financialized peri-urban development has long placed its customers in the path of natural disasters in the search for cheap land and that these risks are becoming significantly less predictable as our climate begins to become less amenable to human life. These challenges have become more pressing and harder to ignore, and it has never been more crucial that the fragmentation of the urban periphery be considered one of the most immediate struggles in adapting to climate change and confronting inequality in urban life.

Just as the driving forces for financialized peri-urban development can be understood and located within the political, social, and economic contexts from which they originate, the problems they create are not a fixed part of life at the edge of the city but a cost of the specific practices of maximally profitable suburbanization on rural land. Throughout this book, I have pointed to the social, ecological, and economic ills born from these practices, but I have also sought to show that solutions at the level of planning, policy, and community already exist, have been espoused by people more qualified and more experienced than myself, and are being called for now more than they have been in recent memory. Aligning the state agenda in favour of local production and consumption, redefining the criteria for productive land to protect it from vicious cycles of collapse and speculation, acting against the financial factors of housing crises and urban sprawl, embracing the unique advantages of peri-urban horticulture, and preparing our cities for a destabilizing climate are all steps that have been taken before and can be taken again. Moving towards a more sustainable and equitable urban future is not only possible but also both practical and necessary.

This rehabilitation of urban life and its relationship to its ecological context is most crucial at the newest frontiers of growing cities. Massively profitable, deeply unequal, extremely inefficient, and uniquely vulnerable to the changing climate, new suburban and exurban developments across the planet are in need of urgent transformation. The manifold crises of the oncoming century which loom over political and academic publications like a horizon-spanning thunderhead are already being felt first and most acutely at the urban periphery, but it is also there where we can find our way to a better future than the one we were given.

# Appendix

## List of processes required for the construction of a suburban home

Note that the list below was designed by a residential construction project manager for their own use, and some entries have been removed or changed in their processual order in order to protect the project manager's competitive concerns.

All descriptions have been directly transliterated from the list, including spelling mistakes, abbreviations and unknown notations.

The first page has the job number, address, house type and size in square metres, salesperson, estimator, and client contact details, with a second page for important notes.

This is then followed by 14 pages containing the following items, each with the job name and then a space for date, time, and contact details.

The process:

1. Surveyor
2. Contact owner
3. Earthworks
4. Soil removal
5. Retaining walls
6. Ensure retaining drainage
7. Eng benching inspection
8. Install site/HazardCo signage
9. Temp fencing erect
10. Portable toilet
11. Rubbish bin
12. Driven piles
13. Pod delivery
14. Drain-layer under slab
15. Sand/metal fill
16. Foundation prep
17. Steelworks/mesh
18. Engineer inspection
19. Water tanks
20. Firefighting tanks
21. Septic system

108. Bobcat site clean
109. Dehumidifier/fans
110. Claim rainy day extension
111. Plumber second draw
112. Electrical walkthrough
113. Electrical rough in
114. Under floor heating pre
115. Alarm pre-wire
116. Gas pipe out
117. Heat pump pipe
118. Fire flue installation
119. Central vac pipe out
120. Shower recess form
121. Insulation batts ordered
122. Insulation batts
123. Hutchinson second inspection
124. Council pre-line
125. Client site meeting
126. Pre-hung doors activated
127. Gib/villa board installation
128. Council post-line inspection

22. Site spoil removal
23. Underground services
24. Council siting inspection
25. Slab pour date
26. Client site meeting
27. Boxing stripped
28. Concrete cut
29. Perimeter drainage
30. Trench and power
31. Temp power
32. Temp water supply
33. Location of power, water, gas
34. Bobcat site clean
35. Make contact with clients
36. Metal delivery
37. Metal spread
38. Slab grind
39. Polished floor first grind
40. Flint coating
41. Frame hardware
42. Frame pre-nail
43. Bulkhead details
44. Builder to site
45. Builder name and number
46. Frame standing
47. Structural beams
48. Lam beam
49. Scaffolding erection
50. Slab trim
51. Safety nets
52. Truss delivery
53. Truss installation
54. Fascia/gutter installation
55. Slab/frame trim
56. Council framing inspection
57. Engineer framing inspection
58. Edge protection
59. Roof installation
60. Gib board measure
61. Garage floor protection
62. Safety nets down
63. Edge protection down
64. Scaffolding lower level (remove?)
65. Client site meeting
66. Eave and gable pack del
67. Bath and niche installation
68. Pergola material pack del

129. Council wet area inspection
130. Pm post-line inspection (LPB to office)
131. Gib stopper
132. Gib cove installation
133. Polished floor final
134. Cardboard floor covering
135. Garage door installation
136. Temp fence removal
137. Wardrobe site measure
138. Waterproofing coatings
139. Kitchen cabinet del
140. Kitchen cabinet installation
141. Kitchen bench template
142. Kitchen bench top installation
143. Kitchen bench top covered
144. Finishing lines delivery
145. Pre-hung doors
146. Finishing carpenter
147. Vanity/shower liners del
148. Vanity shower liners installation
149. White-ware
150. Interior painting
151. Exterior painting touch-ups
152. Special paint finishes
153. Wardrobe shelves (not system)
154. Custom wardrobe site measure
155. Tile delivery
156. Tiler/client site meet
157. Tiled start onsite
158. Tiler completion date
159. Glass shower door/measure
160. Shower liner/doors (framed)
161. Shower doors installation
162. Mirrors installation
163. Notify concrete layer
164. Notify waterfx of CCC date to complete all outstanding works
165. White ware delivery
166. Electrical final fit off
167. Plumbing fit off supplies and delivery
168. Owner supplied fittings
169. Plumber final fit off
170. Gas final fit off
171. Flooring planks prep works

69. Builder install top wrap
70. Builder complete soffits
71. Painter soffit completion
72. Scaffolding complete (not gable)
73. Builder to install RAB board if required
74. Builder wrap/tape
75. Window delivery
76. Window installation complete
77. Site glazing
78. Water penetrations
79. Electrical penetrations
80. Gas penetrations
81. Heat pump penetrations
82. Project manager frame check
83. Cavity sliders to site (255)
84. Rhondo/battens/cavity del (255)
85. Bath/acrylic shower base
86. Wrap/cavity/soffit complete
87. Cavity battens
88. Pre-hung door measure
89. Kitchen site measure
90. Temporary down pipes
91. Temporary water for bricks
92. Ext cladding delivery
93. Cladding install
94. Council wrap inspection
95. Council cladding inspection
96. Soffit/cladding paint
97. Scaffolding removed
98. Bricks delivery
99. Brick hardware
100. Brick hardfill bin
101. Bricks moved around site
102. Bricklayer start
103. Council brick half high
104. Bricks plastered
105. Bricks painted
106. Bricks completion
107. Downpipe install
172. Flooring planks install
173. Skirting board installation over planks
174. Mirror installation
175. Alarm final fit off
176. Central vac fit off
177. Concrete site works
178. Notify drain layer of CCC date for ORG/water catchments/detention/retention
179. Drain final…cest pit heights
180. Septic system commissioned
181. Concrete pour
182. Concrete cut
183. Boxing removed
184. Letterbox installation
185. Gas bottles delivered to site
186. Heat pump/fire fit off
187. Berm soil and seed
188. Site clean
189. Wardrobe shelves installation
190. Carpet installation
191. Fit off hardware to site
192. Carpenter final fit off
193. Wardrobe systems fit out
194. Cleaner first clean
195. Paint tag out
196. Silicon perimeter of boards
197. Client walk-through
198. Paint touch-ups
199. Council CCC inspection
200. Council re-inspection if required
201. Cleaner second clean
202. Water blast/acid wash
203. Check E1 commissioned
204. Toilet pickup
205. Bin pickup
206. Meet client onsite with maintenance supervisor
207. Office to attend handover

A sample of the document, to give an idea of the format when used for job scheduling, is given here.

- Install Site/HazardCo signage:Req......Booked.........By: Txt...... Phone......Email......
- Temp Fencing Erect :Req......Booked......By: Txt......Phone...... Email......
- Portable Toilet :Req.........Booked.........By: Txt......Phone...... Email......
- Rubbish Bin :Req.........Booked.........By: Txt......Phone...... Email......
- Driven Piles :Req.........Booked.........By: Txt......Phone......Email......
- Pod Delivery :Req.........Booked.........By:Txt......Phone...... Email......
- Drain-Layer Under Slab :Req......Booked......By: Txt......Phone...... Email......

# Index

Note: **Bold** page numbers refer to tables, *italic* page numbers refer to figures and page numbers followed by “n” refer to end notes.

For Product Safety Concerns and Information please contact our EU representative GPSR@taylorandfrancis.com
Taylor & Francis Verlag GmbH, Kaufingerstraße 24, 80331 München, Germany

www.ingramcontent.com/pod-product-compliance
Ingram Content Group UK Ltd.
Pitfield, Milton Keynes, MK11 3LW, UK
UKHW021308260926
475702UK00048B/856
*9781032504223*